D1761392

# ANIMAL MACHINES

FSC
www.fsc.org
MIX
Paper from
responsible sources
FSC® C013604

# Animal Machines

Ruth Harrison[‡]

Foreword by Rachel Carson

New contributions from: Marian Stamp Dawkins, John Webster, Bernard E. Rollin, David Fraser and Donald M. Broom

www.cabi.org

**CABI is a trading name of CAB International**

| | |
|---|---|
| CABI | CABI |
| Nosworthy Way | 38 Chauncey Street |
| Wallingford | Suite 1002 |
| Oxfordshire OX10 8DE | Boston, MA 02111 |
| UK | USA |

| | |
|---|---|
| Tel: +44 (0)1491 832111 | Tel: +1 800 552 3083 (toll free) |
| Fax: +44 (0)1491 833508 | Tel: +1 (0)617 395 4051 |
| E-mail: info@cabi.org | E-mail: cabi-nao@cabi.org |
| Website: www.cabi.org | |

First published in 1964 by Vincent Stuart Publishers Ltd.

© J. Harrison and J. Wilson 2013. All rights reserved. No part of this publication may be reproduced in any form or by any means, electronically, mechanically, by photocopying, recording or otherwise, without the prior permission of the copyright owners.

A catalogue record for this book is available from the British Library, London, UK.

**Library of Congress Cataloging-in-Publication Data**

Harrison, Ruth, 1920-2000.
    Animal machines / Ruth Harrison; foreword by Rachel Carson; new commentaries by Marian Stamp Dawkins … [et al.].
        p. cm.
    First published: London: Vincent Stuart Publishers, 1964.
    Includes bibliographical references and index.
    ISBN 978-1-78064-284-0 (alk. paper)
    1. Factory farms. 2. Animal welfare. 3. Animal industry. 4. Livestock. I. Carson, Rachel, 1907-1964. II. Dawkins, Marian Stamp. III. Title.
    SF140.L58H38 2013
    636--dc23
                                    2012050255

ISBN-13: 978 1 78064 284 0

Commissioning editor: Sarah Hulbert
Editorial assistant: Alexandra Lainsbury
Production editor: Shankari Wilford

Typeset by SPi, Pondicherry, India
Printed and bound in the UK by CPI Group (UK) Ltd, Croydon, CR0 4YY.

# Contents

1   **Why We Still Need to Read** *Animal Machines*        1
    *Marian Stamp Dawkins*

2   **Ruth Harrison – Tribute To An Inspirational Friend**   5
    *John Webster*

3   *Animal Machines* **– Prophecy and Philosophy**         10
    *Bernard E. Rollin*

4   **Ruth Harrison – A Tribute**                           17
    *David Fraser*

5   **Ruth Harrison's Later Writings and Animal Welfare Work**   21
    *Donald M. Broom*

**Foreword to 1964 Edition by Rachel Carson**               31

**Acknowledgements**                                        33

  I. **Introduction**                                       35

 II. **Broiler Chickens**                                   41

III. **Poultry Packing Stations**                           56

IV. **Battery Birds**                                       65

 V. **Veal Calves**                                         85

VI.  Other Intensive Units                                                    106

     The New Factory Farming – A Pictorial Summary               114

VII.  The Basis of Quality                                                    139

VIII.  Quantity Versus Quality                                              148

IX.  Cruelty and Legislation                                                173

X.  Conclusion                                                                    194

Bibliography                                                                        205

Index                                                                                  207

# Why We Still Need to Read *Animal Machines*

**1**

**Marian Stamp Dawkins**
*University of Oxford, UK*

Much has changed about farm animal welfare since the publication of *Animal Machines* in 1964, and much of that change can be traced back directly to this pioneering book. Fifty years ago, Ruth Harrison called for the complete abolition of the battery cages that increasingly were being used to keep laying hens. These remained legal and widespread for many years but are now banned in many parts of the world. Sweden and Switzerland were the first to take action but, on 1 January 2012, the European Union banned them throughout its 27 member states. The USA still has the majority of its laying birds in cages but the State of California banned them unilaterally in 2008, and a bill currently (2013) going through Congress will impose a complete federal ban. As John Webster describes in his personal tribute (Chapter 2), Ruth's influence can be detected at every stage in these and other changes, from the setting up of the Farm Animal Welfare Council in the UK, to her dogged pursuit of changes to animal law through various European committees. Pigs, veal calves, geese, broiler chickens and other food animals are, at least in some parts of the world, similarly treated very differently now from how they were in 1964.

In addition to the impact it has had on our current laws and regulations for farm animal welfare, *Animal Machines* also spoke directly to the hearts and minds of the general public. It was one of those books that changed the way people thought forever. Before *Animal Machines*, the public was largely unaware of the innovations that were taking place in agriculture to make it more 'intensive' and efficient, usually by bringing animals indoors. After *Animal Machines*, and the publicity that it gave rise to, it was no longer possible to plead ignorance of such matters.

© J. Harrison and J. Wilson 2013. *Animal Machines* (Ruth Harrison)     1

Ruth's book had started something that ricocheted around the world and affected the philosophical foundations of ethical thinking about animals, as Bernie Rollin explains in Chapter 3. It also led eventually to animal welfare itself becoming rooted much more firmly in scientific evidence, as both David Fraser (Chapter 4) and Donald Broom (Chapter 5) describe from different viewpoints. Animal welfare went from being the concern of a sentimental few to a matter of great importance to the public at large. It was once 'beyond the pale' of scientific study and then came in from the cold to become the flourishing science of animal welfare that it is today. Ruth herself funded scientific research into animal welfare at a time when little or no other funding was available. Now, 'improving animal welfare' is one of the key ways a scientist can demonstrate the importance and impact of their work.

In one sense, then, *Animal Machines* is a book of its time, to be read as describing conditions as they were 50 years ago, not as they are now. A perfectly good reason for reprinting it thus would be out of respect for its landmark significance in the history of animal welfare and to remind us how far we have come since then. But in another sense, *Animal Machines* is also a book of our own time, for the simple reason that Ruth's work is not yet done and we are still in need of her skills as a mover and shaker and changer of received opinions.

We live at a time of rising meat consumption across the world. As people become richer, they tend to want more meat, with the result that more and more animals are being farmed for food. At the same time, the human population is rising rapidly, with a projected 9 billion humans on the planet by 2050 (Godfray *et al.*, 2010). To meet the problem of feeding so many people, the United Nations and governments around the world are calling for livestock production to become more efficient and more intensive (Steinfeld *et al.*, 2006; The Government Office for Science, 2011). The watchword is now that agriculture must become 'sustainably intensive', raising fears that the improvements in animal welfare we have seen over the past 50 years might become lost, or even reversed, unless we are as vigilant and perceptive now as Ruth was then.

On the very first page of *Animal Machines*, Ruth uses a succinct description of intensive animal production: 'Rapid turnover, high-density stocking, a high degree of mechanization, a low labour requirement, and efficient conversion of food into saleable products.'

She then went on to describe the effects that such intensive methods were having on chickens kept for meat and eggs, calves, dairy cattle, pigs and rabbits, and it was those effects that caused such a public outcry. The definition of intensive farming that she gave would still stand today. The challenge for us is how to combine the need for 'efficient' food production with maintaining and improving standards of animal welfare. Another way of putting this is that if 'sustainable intensification' is what the world wants, then we need to take that wobbly word 'sustainable' and give it enough backbone to make sure that it stands for high standards of animal welfare and safeguards it as one of its cornerstone priorities. It is here that *Animal Machines* comes into its own as directly relevant to the present time and the dilemmas that face us now.

One of the most striking things about Ruth's book is the diversity of arguments she uses to support better standards of animal welfare. Of course she emphasizes the importance of avoiding cruelty and improving the lives of animals, but she also spends a lot of time explaining how good animal welfare is better for humans, too. There is one whole chapter on food quality for humans; another dealing with the ecological impact on human lives; and the final chapter is almost entirely about money and the financial benefits of providing good animal welfare. The book is extraordinarily pragmatic. Ruth acknowledges that not everyone will agree with her and then sets about trying to convince those who do not with every argument at her disposal – using their financial self-interest, their concern for their own health, the quality of the lives they would like to lead or anything else that might possibly bring them round. In this she is ahead of her time. It is only now that we are beginning to appreciate the importance of a fully integrated approach to animal welfare in which other concerns, such as environmental impact and human health, are taken alongside each other (Dawkins, 2012). We still have much to learn from what Ruth Harrison was already saying all those years ago.

Ruth had the vision to ask Rachel Carson, author of another landmark book, *Silent Spring* (1962), to write the Foreword to *Animal Machines*, thus emphasizing further the connection between environmental impact and animal welfare. Rachel Carson was evidently very affected by Ruth's book. 'Wherever it is read, it will certainly provoke feelings of dismay, revulsion and outrage', she wrote (p. viii). She also recognized that Ruth, with her combination of 'patient scholarship and steadfast courage' (p. vii), had been able to write such a powerful book precisely because she had drawn together a set of diverse arguments (humanitarian, threat to human health, economic) to make her case.

Even now, the full potential of such an integrated approach to animal welfare has not been totally realized and we have not taken on board the many different ways Ruth Harrison used to bring about change – emphasizing scientific evidence, looking broadly at the societal impact of the way we treat animals, listening to other points of view and then trying systematically to find ways of bringing a whole range of people around to her way of thinking. She sought information and wanted everyone to have access to it. She persuaded, she argued and she never, never gave up. She 'made a difference', to use a current phrase, and anyone wanting to make a similar difference today would do well to start by reading *Animal Machines*. The differences between then and now show what can be achieved. The similarities are an inspiration for what we still have to do today.

# References

Carson, R. (1962) *Silent Spring*. Published in the UK by Hamish Hamilton, London (1963). This is the edition RH cites.

Dawkins, M.S. (2012) *Why Animals Matter: Animal Consciousness, Animal Welfare and Human Well-being*. Oxford University Press, Oxford, UK.

Godfray, H.C.J., Beddington, J.R., Crute, I.R., Haddad, L., Lawrence, D., Muir, J.F., *et al.* (2010) Food security: the challenge of feeding 9 billion people. *Science* 327, 812–817.

Government Office for Science, The (2011) *Foresight: The Future of Food and Farming. Challenge and Choices for Global Sustainability*. Final Project Report. The Government Office for Science, London.

Harrison, R. (1964) *Animal Machines. The New Factory Farming Industry*. Vincent Stuart, London.

Steinfeld, H., Gerber, P., Wassener, T., Castel, V., Rosales, M. and de Haan, C. (2006) *Livestock's Long Shadow. Environmental Issues and Options*. Food and Agricultural Organization of the United Nations.

# Ruth Harrison – Tribute To An Inspirational Friend

**2**

**John Webster**
*University of Bristol, UK*

The British nation likes to think that it is kind to animals. Since 1822, there have been laws for the protection of farm animals in the UK. Before then, animals were defined as property and had no legal protection in their own right. In 1822, the UK parliament passed 'Martin's Act' to prevent the cruel and improper treatment of cattle, making this the first parliamentary legislation for animal welfare in the world. The most robust single piece of legislation was the Protection of Animals Act 1911, which made it an offence to 'cause unnecessary suffering by doing, or omitting to do, any act'. The implementation and interpretation of the law are matters for the judiciary. However, laws are written by politicians and, in a democracy, politics is expected to adapt and evolve to reflect the changing perceptions of people for a just society. This is often interpreted as that which is seen as just by the 'reasonable man'.

It is nevertheless a happy circumstance that (to quote Hart, 1961) 'It cannot seriously be disputed that the development of the law... has been profoundly influenced both by conventional morality and ideals of particular social groups, and also by forms of enlightened moral criticism urged by individuals whose moral horizon has transcended the morality currently accepted.' So often, the enlightened moral critics have not been reasonable men but remarkable women, among such notables being Elizabeth Fry, Emmeline Pankhurst, Rachel Carson and Ruth Harrison. In her massively influential book, *Animal Machines* (1964), Ruth wrote, 'If one person is unkind to one animal it is considered as cruelty but where a lot of people are unkind to a lot of animals, especially in the name of commerce, the cruelty is condoned and, once large sums of money are at stake, will be defended to the last by otherwise intelligent people.' This single sentence laid bare the mindlessness or wilful hypocrisy necessary to justify so much of intensive farming. In effect, both producers and consumers (i.e. all of us) were content to offer a life worth living only to those animals we got to know as individuals; like our pets. It was Ruth who pointed out the absurdity of the UK's Protection of Birds Act that had been passed in 1954 and required any caged bird to be given enough space to flap its wings but then stated *provided this subsection shall not apply to poultry'* (the italics were Ruth's own; *Animal Machines*, p. 153).

There was little in her upbringing to presage the sustained sense of outrage and demand for the truth that led her to research and write *Animal Machines*. Ruth was a vegetarian (not vegan). However, prior to 1961, she had enjoyed a productive, artistic and essentially urban life but (so far as I can gather) not displayed any special sentimental attachment to animals. The starting gun for her crusade was apparently a single leaflet from the 'Crusade Against All Cruelty to Animals' that drew her attention to the plight of animals raised for food, such as veal calves, broiler chickens and caged laying hens. She was moved to action not primarily, she said, from a love of animals but a burning sense of injustice to those whom she recognized as sentient creatures that deserved to live rather than simply exist before they were slaughtered for our satisfaction.

Today, *Animal Machines*, the book, should be read the way one reads Aristotle or the Bible: with great respect for its power and insight, but not to be taken as gospel. Much of what she describes has changed, in a small part through new legislation and in a greater part through the direct power of the people, demonstrated most conspicuously by the huge increase in demand for free range eggs. Some of her conclusions, reasonable at the time, need to be reinterpreted in the light of new understanding of the physiological and behavioural needs of farm animals, and a few are simply wrong. (I leave these for you to discover.) Nevertheless, the evolution of major improvements in farm animal welfare for pigs, calves and chickens through legislation in the UK and European Union, the state-by-state legislation to ban sow stalls in the USA, the development of high welfare schemes like Freedom Foods and the Global Animal Partnership, and the massive increase in funding for the pursuit and application of animal welfare science (e.g. the pan-European Welfare Quality® programme) can all can be traced back, like mitochondrial DNA (the female line), to the common ancestor, namely Ruth herself.

As I reread *Animal Machines*, I am struck once again by its power. Ruth did not hesitate to present verbal and visual images of the worst of factory farming in order to maximize the emotional impact of her argument. She was fully aware of the power of the image. The picture of the battered hen in the battery cage has done more to influence public opinion than a thousand diligent scientific studies of the welfare of the laying hen. However, she was always meticulously careful to back up her images with evidence. Most of this was obtained by visiting the factory farms and talking with the producers, many of whose written and verbal explanations and justifications were quoted directly and in detail. There was little welfare science in the book, for the good reason that there was very little welfare science around at the time. Animal science then was directed almost exclusively towards increased efficiency of production – and I can speak as a scientist who was beginning his research career in the year that *Animal Machines* was published. At the end, she does not condemn the industry outright. Having set down the evidence, quite fairly, she asks you to make up your own mind: which is easy.

I first met Ruth when I was appointed to the Farm Animal Welfare Advisory Committee, the predecessor of the FAWC, the Farm Animal Welfare Council, in (I think) 1979. I had recently moved from a full-time research job at the Rowett Research Institute in Aberdeen to take up the Chair of Animal Husbandry at the University of Bristol Veterinary School. While at the Rowett, I was asked to carry out an independent review of research into mineral requirements (mainly iron and copper) for the production of white veal – the primary aim being to ensure the quality of the meat without serious compromise to productive efficiency. This was my first exposure to animal production at its worst. I concluded that the solution to the veal calf problem was not just a matter of fine-tuning iron supply to avoid clinical anaemia. Everything was wrong! Lack of fibre in the diet caused chronic ill health resulting from digestive disorders, and these predisposed to secondary pneumonia. The accommodation created chronic physical and thermal discomfort. The animals were denied natural oral, comfort and social behaviour. Lacking experience of the normal sights and sounds of farm activities, they panicked at the slightest alarm. Reread the last four sentences and you will see that I have described abuse of all the Five Freedoms.

On arrival at Bristol, my first application for a research grant from the Agricultural Research Council (ARC, now BBSRC) was for work designed to improve the health and welfare of veal calves through the development of more humane, although comparably efficient, alternative husbandry systems. At the time, the ARC rejected my application on the grounds that 'because insufficient is known about this subject at the present time we feel this research is premature' (sic). Thus, when I first met Ruth, we discovered that we had a lot of outrage in common. The outcome of this was that the very first research funding for what has now become the large and internationally respected University of Bristol Centre for Animal Welfare and Behaviour came from the Farm Animal Care Trust, as established by Ruth. This enabled Claire Saville (now Weekes) and I to begin a programme of research designed to address the major welfare problems of veal calves. The main conclusions of this work and that done later by David Welchman with the support of the Animal Health Trust (funded by another remarkable lady, Mrs Allen) were transcribed almost word for word into UK (and later European Union) legislation defining minimal standards for the feeding and housing of calves reared for veal.

During her time on the FAWC, Ruth was always passionate in the pursuit of justice but equally open to reasoned argument and to new knowledge emerging from developments in science and practice. However, she retained her scepticism in regard to comforting assurances. On more than one occasion, she displayed the true courage of those early physiologists by testing things out on herself. Ruth submitted herself to procedures for carbon dioxide stunning and electro-immobilization, both promoted as humane. The first she pronounced terrifying, the second excruciating. We believed her and acted accordingly.

The relentless vigour with which Ruth pursued her campaign was softened, and thereby made more effective, by her charm (she had a splendidly earthy laugh) and her willingness to listen politely to counterarguments. However, rarely would she accept these counter-arguments at face value, whether from the industry or scientists. She would not dismiss them outright with an emotional sweep of hand but, confident in the fact that she was as well informed as her interlocutor, pick out specific errors and shortcomings and insist on review and remedy. When we were drafting reports for the FAWC, getting towards the end of the day and looking forward to the first drink or going home, she would regularly draw attention to areas where she was not satisfied and where she thought we could and should do better. This could be exasperating, but since, in my opinion, she was nearly always right, I had to admit that she was magnificent.

It is a matter of historical record that the publication of *Animal Machines* was the stimulus for the Brambell Report (1965) that put farm animal welfare as a top priority on the political agenda. Much more than that, it was the torch that lit the flame of the farm animal welfare movement that has, in the past 50 years (and especially the last 10), brought about real and big improvements in the welfare of farm animals. This has come about in part through legislation for improved minimum standards for animals in the intensive systems that caused her most concern: confined veal calves, sows and laying hens. However, the major force for change has been and will continue to be through increased public awareness of the problems going on behind closed doors, demand for improvements and action to reward more humane husbandry through the purchase of quality assured and quality controlled high welfare food from animals. We still have a long way to go, but the extent and accelerating pace of improvement has far exceeded my expectations. It is a great shame that Ruth died just before the pace of improvement really started to quicken. However, her legacy is immortal.

# References

Brambell, F.W.R. (1965) *Report of Technical Committee to Enquire into the Welfare of Animals Kept Under Intensive Husbandry Systems*, Cmnd. 2836. HMSO, London.

Harrison, R. (1964) *Animal Machines: The New Factory Farming Industry*. Vincent Stuart, London.

Hart, H.L.A. (1961) *The Concept of Law*. Clarendon Press, Oxford, UK.

# Animal Machines – Prophecy and Philosophy

**3**

**Bernard E. Rollin**
*Colorado State University, USA*

The word 'prophet' is often misused. Today, it may connote a supernatural crystal gazer or tea-leaf reader. Yet historically, back to biblical times, it meant a person who was aware clearly and vividly of social moments that others ignored or discounted, as when Jeremiah reasonably prophesied the destruction of Israel if it continued to antagonize the superpower Babylon.

It is in the latter sense that Ruth Harrison was and is unquestionably a prophet in the true meaning of the word. As it happens, I was in Britain as a young student in 1964 when *Animal Machines* was published. At no other point in my career have I seen a single book on its own so electrify a whole society. The book occasioned a great flood of publications in the mass media, including an excellent series by Elspeth Huxley that I read with a mixture of fascination and horror, and which synergistically reinforced the Harrison book. I was not yet a student of animal ethical issues, specializing instead in the quite rarified matters of David Hume and Scottish Common Sense Philosophy while at the University of Edinburgh. Yet, even then, it was clear to me that Harrison's writing was in the best tradition of Hume in acknowledging animal suffering as a common-sense truism, and a matter of common decency, an insight that would, in the 1970s, inform and shape my career in animal ethics for the next four decades.

*Animal Machines* represented a unique blend of powerful emotional impact, unforgettable prose and cutting-edge scientific knowledge. Virtually every issue pertaining to the untoward effects of the industrialization of agriculture, from animal welfare issues, to human health concerns, to environmental issues, was expressed strikingly in the book.

© J. Harrison and J. Wilson 2013. *Animal Machines* (Ruth Harrison)

And not only did the book glaringly illuminate this multiplicity of issues – Harrison was extraordinarily adept at another necessary condition for functioning as a prophet. Her prose style was such that it could touch the ordinary people in the British public on both an intellectual and a visceral level, so powerfully that in the wake of the book, the British government was forced to charter a commission of enquiry into the lives of animals raised under intensive confinement conditions. The commission's report (Brambell, 1965) codified what has subsequently become known as the Five Freedoms: conceptually, a series of minimal moral rights for farm animals, later adopted by the Farm Animal Welfare Advisory Committee and eventually, in 1979, in the Farm Animal Welfare Council (FAWC). The original Brambell Report stated: 'Farm animals should have the freedom to stand up, lie down, turn around, groom themselves and stretch their limbs.' In the current language of the FAWC, the Five Freedoms are described as follows:

## Five Freedoms

The welfare of an animal includes its physical and mental state and we consider that good animal welfare implies both fitness and a sense of well-being. Any animal kept by humans must, at least, be protected from unnecessary suffering.

We believe that an animal's welfare, whether on farm, in transit, at market or at a place of slaughter, should be considered in terms of 'five freedoms'. These freedoms define ideal states rather than standards for acceptable welfare. They form a logical and comprehensive framework for the analysis of welfare within any system, together with the steps and compromises necessary to safeguard and improve welfare within the proper constraints of an effective livestock industry.

1. **Freedom from Hunger and Thirst** – by ready access to fresh water and a diet to maintain full health and vigour.
2. **Freedom from Discomfort** – by providing an appropriate environment including shelter and a comfortable resting area.
3. **Freedom from Pain, Injury or Disease** – by prevention or rapid diagnosis and treatment.
4. **Freedom to Express Normal Behaviour** – by providing sufficient space, proper facilities and company of the animal's own kind.
5. **Freedom from Fear and Distress** – by ensuring conditions and treatment which avoid mental suffering.

Unfortunately, the Brambell Report did not enjoy any regulatory authority. Despite this major impediment, the report served as a beacon illuminating the issues raised by factory farming and eventually did acquire regulatory teeth in European Union legislation and in legislation adopted by numerous individual European countries, most notably 1988 Swedish law severely restricting the confinement of agricultural animals, largely developed by iconic Swedish author, Astrid Lindgren, who is often described in Sweden as 'everyone's grandmother', with the help of Swedish farm animal veterinarian, Christina Forslund, who showed Lindgren what had happened to farm animals in Sweden. Described by the *New York Times* as 'a bill of rights for farm animals', the law sailed through the Swedish parliament virtually unopposed. Once again, the prophetic vision of Ruth Harrison is clearly evidenced in the historical progression described and still ramifies in US state referenda abolishing battery cages, sow stalls or gestation crates and veal crates, spearheaded by the Humane Society of the United States.

When I began to write about animal ethics in the mid-1970s, the influence of the chain of thought just described, beginning with Ruth Harrison and moving through the Brambell Commission, was dominant in my thinking. In particular, two major links in this chain were critical to my approach – enlisting public support and respecting animals' physical and psychological natures. By then, Peter Singer, working out of England and necessarily influenced by Harrison's work, had written his seminal work, *Animal Liberation* (1975), self-described as a 'new ethic for animals', based in the grand British utilitarian tradition tracing back directly to the work of Jeremy Bentham. While I admired Singer's arguments greatly, I was not satisfied with them, for several reasons. First of all, the thrust of Singer's reasoning was to abolish animal use that did not seem morally justifiable in utilitarian terms. Second, at least as I understood Singer's approach, it was based on maximizing pleasure and minimizing pain. While such an approach was commendable, it seemed to me that simply measuring pleasure and pain along a single axis was unsatisfactory. I identified many ways in which human uses harmed animals, yet these did not seem to fit under the notion of 'pain'. In particular, I recognized that not every sort of harm humans inflict on animals counts as *pain*, in any ordinary sense of the word, unless the notion of pain is stretched so broadly as to cover all sorts of misery not usually classified under the rubric of pain. Such states as loneliness, boredom, fear, stimulational deprivation, inability to exercise, separation from offspring or parents, impoverished diet, inability to forage, hunt or otherwise secure one's own food, and myriad others, certainly harm animals, and had been identified as doing so by Ruth Harrison, yet do not without great artificiality lend themselves to being arrayed along an axis of pleasure and pain. Second, and equally problematic, while Harrison had identified components of animal use eliciting righteous indignation from the general public, I did not see the vast majority of that public being willing to forgo foods of animal origin or, as regards the issue of invasive research on animals, to forgo the benefits that animal experimentation could yield for human health.

As the great radical activist, Henry Spira, said to me on many occasions, there was never any social/ethical revolution in the history of the USA that was not gradual and incremental. This was even true of the civil rights movement we both lived through – why would anyone think that moral enfranchisement of animals would be any different? As I continued to develop my own ideas, I crafted my approach based on the influence of Harrison, Plato and Aristotle. Harrison, in her writings, as well as in her considerable influence on policy in Europe, had stressed the importance of respecting animals' psychological and biological needs and natures, as had the Brambell Commission. Furthermore, the notion of animal *nature* was ensconced solidly in ordinary common sense, though rejected by contemporary biological science. The vast majority of ordinary people would find the notion of 'the pigness of a pig; the cowness of a cow; the dogness of a dog' conceptually unproblematic and see the violation of such natures in the manner done by confinement agriculture as monstrous. (When I served on the Pew Commission on Industrial Farm Animal Production with some 15 experts on matters related to food of animal origin and confinement agriculture, many commissioners, who had never before seen a sow barn based in high confinement gestation crates, left the barn in tears.) As a person who, for much of my career, had taught the history of philosophy, I was able to draw on Aristotle's concept of *telos*, animal nature, as the root concept behind my account of moral obligations to animals, a notion that was as persuasive to the thousands of cowboys to whom I have lectured on animal ethics as it was to consumers.

The final link in my approach to animal ethics came from Plato and was illuminated clearly in Ruth Harrison's work. Throughout Plato's work, he had stressed that, when dealing with ethics in adults, one could not *teach*, one needed to *remind*. Having lived through the civil rights era, I realized that Martin Luther King and Lyndon Johnson had acted in accord with Plato's admonition to remind rather than teach. Their successful appeal to the US public was not based in creating new moral principles for the treatment of black Americans. It was rather their *reminding* citizens of their commitment to the notion that all people should be treated equally and that black people were, indeed, people.

To supplement and elucidate Plato's notion of reminding versus teaching, I created my own metaphorical explanation of the strategy I deployed in terms of martial arts. There are two distinct and antithetical approaches to hand-to-hand combat. One is a sumo approach, wherein one exerts one's force against the force of one's opponent in the manner of offensive versus defensive linemen in football. This is a viable approach if you and your opponent are of equal size and strength; ideally, you are larger. It is a recipe, however, for certain defeat if you are fighting someone of significantly superior size and strength. In such a case, one is far better advised to use an opponent's strength against that opponent, judo, so that you redirect that strength to unbalance the opponent, or to throw them. The logic similarly obtains in ethical debate. Particularly if one is arguing against a more powerful opponent, one fares far better by showing that opponent that your ethical position is implicit in their own ethical assumptions, albeit in a hitherto unnoticed way, rather than attempting to force your position on them.

This, in turn, brought me to a new realization regarding animal ethics. If, as appeared to be the case, Western society was moving steadily towards greater moral concern and moral status for animals, it would not do so by creating a totally new ethic for animals *ex nihilo*. Rather, it would look to our extant social ethic for the treatment of humans and export it, *mutatis mutandis*, appropriately modified, to the treatment of animals.

While such a move is not always possible, it is far easier to accomplish than one would think. After all, a good deal of societal education is devoted to assuring that we all grow up with the same social ethical skeleton or core. Indeed, if we did not in a significant way share much the same foundational ethical beliefs, it would be difficult for society to function in a non-anarchistic way. Given the extent to which we all share an ethical skeleton, which society works very hard to instil in all of us, beginning in childhood, would it not be far more fruitful to attempt to deduce the logical extension of that ethic, were society to wish to apply that ethic to animals, than to create a new ethic from whole cloth, an ethic that might well lack a point of contact with what most people already believe ethically?

At the same time, recall that Western society had gone through almost 50 years of extending its moral categories for *humans* to people who were morally ignored or invisible – women, minorities, the handicapped, children, citizens of the Third World. As we noted earlier, new and viable ethics do not emerge *ex nihilo*. So, a plausible and obvious move is for society to continue in its tendency and *attempt to extend the moral machinery it has developed for dealing with people, appropriately modified, to animals*. And this is precisely what has occurred. Society has taken elements of the moral categories it uses for assessing the treatment of people and is in the process of modifying these concepts to make them appropriate for dealing with new issues in the treatment of animals, especially their use in science and confinement agriculture.

What aspect of our ethic for people is being so extended? One that is, in fact, quite applicable to animal use is the fundamental problem of weighing the interests of the individual against those of the general welfare. Different societies have provided different answers to this problem. Totalitarian societies opt to devote little concern to the individual, favouring instead the state, or whatever their version of the general welfare may be. At the other extreme, at least some anarchical groups such as communes give primacy to the individual and very little concern to the group – hence, they tend to enjoy only transient existence. In our society, however, a balance is struck. Although most of our decisions are made to the benefit of the general welfare, fences are built around individuals to protect their fundamental interests from being sacrificed to the majority. Thus, we protect individuals from being silenced, even if the majority disapproves of what they say; we protect individuals from having their property seized without recompense, even if such seizure benefits the general welfare; we protect individuals from torture, even if they have planted a bomb in an elementary school and refuse to divulge its location. We protect those interests of the individual that we consider essential to being human, to *human nature*, from being submerged, even by the common good. Those moral/legal fences that so protect the individual human are called *rights* and are based in plausible assumptions regarding what is essential to being human.

It is this notion to which society in general is looking in order to generate the new moral notions necessary to talking about the treatment of animals as regards their natures in today's world, where cruelty is not the major problem, as Ruth Harrison argues so beautifully in the last chapter of her book, but where such laudable, general human welfare goals as efficiency, productivity, knowledge, medical progress and product safety are responsible for the vast majority of animal suffering. People in society are seeking to 'build fences' around animals, to protect them and their interests and natures from being totally submerged for the sake of the general welfare, and are trying to accomplish this goal by going to the legislature. In traditional agriculture based in good husbandry or care to assure productivity, this occurs automatically; in industrialized agriculture, where it is no longer automatic, people wish to see it legislated.

It is necessary to stress here certain things that this ethic, in its mainstream version, is *not* and does not attempt to be. As a mainstream movement, it does not try to give human rights to animals. Since animals do not have the same natures and interests flowing from these natures as humans do, human rights do not fit animals. Animals do not have basic natures that demand speech, religion or property; thus, according them these rights would be absurd. On the other hand, animals have natures of their own and interests that flow from these natures, and the thwarting of these interests presumably matters to animals as much as the thwarting of speech matters to humans. The agenda is not, for mainstream society, making animals have the same rights as people. It is rather preserving the common-sense insight that 'fish gotta swim and birds gotta fly', and suffer if they do not.

This new ethic is *conservative*, not radical, harking back to the husbandry-based animal use that necessitated and thus entailed respect for the animals' natures. It is based on the insight that what we do to animals *matters* to them, just as what we do to humans matters to them, and that consequently we should respect that mattering in our treatment and use of animals as we do in our treatment and use of humans. *And since respect for animal nature is no longer automatic, as it was in traditional husbandry agriculture, society is demanding that it be encoded in law.* Significantly, in 2004, in the USA, no fewer than 2100 bills pertaining to animal welfare were proposed in US state legislatures. Over 90 law schools now teach animal law.

With regard to animal agriculture, the pastoral images of animals grazing on pasture and moving freely are iconic. As the 23rd Psalm indicates, people who consume animals wish to see animals live decent lives, not lives of pain, distress and frustration. It is for this reason in part that industrial agriculture conceals the reality of its practices from a naïve public.

The final sense in which we can view Ruth Harrison as prophetic is thus in her ability to see where ordinary citizens stand regarding ethical obligations to animals and 'reminding them' of these obligations. If, as I argue at the beginning of this essay, a prophet points out clear social elements and issues otherwise unheeded, and makes people attend to them, then surely Ruth Harrison is the prophet of animal ethics, as certainly as Martin Luther King is the prophet of civil rights.

## References

Brambell, F.W.R. (1965) *Report of Technical Committee to Enquire into the Welfare of Animals Kept Under Intensive Husbandry Systems*, Cmnd 2836. HMSO, London.

Farm Animal Welfare Council (1979) (http://www.defra.gov.uk/fawc/about/five-freedoms, accessed December 2012).

Harrison, R. (1964) *Animal Machines: The New Factory Farming Industry*. Vincent Stuart, London.

Singer, P. (1975) *Animal Liberation: A New Ethics for Our Treatment of Animals*. Jonathan Cape, London.

# Ruth Harrison – A Tribute

**4**

**David Fraser**
*University of British Columbia, Canada*

In Edinburgh during the early 1970s, David Wood-Gush led a programme of animal welfare research from a small office cluttered with papers filed in what he called the deep-litter system. There was room for two visitors to sit and talk, but any third person had to sit on a chair behind a filing cabinet, hidden from half the room. Ruth Harrison took great interest in our research on farm animal welfare and on one of her visits she was occupying that chair during a discussion with David, Ian Duncan and myself. After some time, a colleague – an agriculturalist who seemed perplexed by all the fuss about farm animal welfare – appeared at the door to see whether our conversation had ended.

'Ah, I see you've got rid of the good lady,' said the colleague.

'No,' came the voice of Ruth Harrison from behind the filing cabinet, 'I'm still here.'

And indeed she is. In fact, it is impossible to tell the history of farm animal welfare and animal welfare science without recounting the influence of Ruth Harrison and *Animal Machines* (e.g. Fraser, 2008; Woods, 2012). *Animal Machines*, as is well known, triggered an astonishing level of public concern in 1964 when it was first published and simultaneously serialized in a major British newspaper. The British government was reportedly inundated by citizens concerned that they were supporting institutionalized cruelty through their grocery purchases and poisoning themselves with unsafe food at the same time. In response to the furor, the government quickly appointed a committee, chaired by the eminent Prof F.W. Rogers Brambell, to investigate 'the welfare of animals kept under intensive livestock husbandry systems'. In view of what the committee

© J. Harrison and J. Wilson 2013. *Animal Machines* (Ruth Harrison)

described as 'the extent of public disquiet' on the subject, they worked with a sense of urgency (Brambell, 1965).

The Brambell Committee report was published the year after *Animal Machines*. As well as the usual recommendations and recounting of evidence, the report included a remarkable appendix written by one of the members, William Thorpe, a gentlemanly Quaker pacifist known for his scientific studies of birdsong. The appendix was a thoughtful essay on 'the assessment of pain and distress in animals'. It outlined veterinary studies of disease and injury, physiological indicators of stress, behavioural indicators of pain and discomfort, studies of motivation thwarted in confinement, research on the intelligence and cognitive powers of animals, studies of animals' capacity to develop a learned fear of humans and the significance of the preferences that animals showed for different environments (Thorpe, 1965). The essay was virtually an agenda for the scientific study of animal welfare that would unfold over the next half century. Under Thorpe's influence, the committee called for a new type of research to be done, partly to shed light on how the welfare of animals was influenced by different production methods and partly to allow animal agriculture to become more efficient by better matching the needs and nature of the animals.

The committee's call for research effectively launched the field of animal welfare science. One result was the expansion of the Ethology Section of the Poultry Research Centre in Edinburgh, which was led by David Wood-Gush and came to include such influential figures as Ian Duncan, Barry Hughes, Michael Gentle, John Savoury and Brian Jones, all of whom made significant contributions to animal welfare science and trained a generation of graduate students who are leaders in the field today. My own position, as the first scientist hired to study the welfare of farm animals at the (then) Edinburgh School of Agriculture, was a direct result of *Animal Machines* and the Brambell Committee and was followed, some years later, by the establishment of a large and continuing research group in Edinburgh. And the creation and funding of other animal welfare research in the UK undoubtedly resulted from the influence of the book and the committee report.

But Ruth Harrison did not simply trigger the events that led to animal welfare research – she maintained an active interest in animal welfare science and reform throughout her life. She was a long-standing member of the UK's influential Farm Animal Welfare Council, where she served alongside scientists, agriculturalists and veterinarians. Apart from such official involvement, she remained an informed critic of animal production, as I had several occasions to witness. One time, when we were both attending a conference in Maryland, USA, she discovered that North America's first experimental unit with robotic milking of dairy cows was located within driving distance of the meeting and she arranged a visit to the facility and kindly included me in the trip. Another time, while I was working in Canada, she telephoned from England to say that an early mechanical chicken catcher – a machine that had the potential to replace stressful manual catching of chickens – was being demonstrated to potential buyers in the southern USA and she seemed unable to understand why I, as a government scientist employed to do research on the welfare of pigs, was not at liberty to down tools and fly to the demonstration. In 1987, when she gave a special address at a bioethics conference in Montreal, Canada, she extended her trip to tour our research unit in Ottawa. I had hosted numerous animal welfare advocates over the years, but Ruth Harrison was one who made me feel that I was dealing with an informed colleague. (She was, however, slightly horrified that we expected even 'ladies' with freshly styled hair to shower before going into the high-health pig unit.)

Since Ruth Harrison, there have been many other critics of people's treatment of animals. Some have been more accomplished writers; some have rooted their arguments more strongly in ethical theory; some have been more comprehensive in their approach rather than focusing only on farm animals; but none has equalled Ruth Harrison in the sheer effect she had. Why is this?

There are many reasons, of course: the novelty of her message back in 1964, her obvious sincerity and the cultural resonance, in a country whose art and literature still reflected deeply divided views of industrialization, of the claim that animal agriculture was becoming an industrial activity. But one of the reasons for Ruth Harrison's impact was her respect for factual information and her informed engagement with the real-life problems of animal use.

In this respect, Ruth Harrison's example is needed now more than ever before. All too often, critics today simply repeat claims that are so out of date or so ill-informed that they are easy for farmers and other animal users to dismiss as merely mistaken; and animal users, concluding that their critics are ignorant, respond with facile public relations (Fraser, 2001, 2012). In this battle of rhetoric and counter-rhetoric, it is hard to find the informed critics whose knowledge allows them to engage in the process of reform rather than simply criticizing from the sidelines and alienating those who actually work with animals.

I am sure that others will draw different lessons from Ruth Harrison's determination, her persuasiveness and her lifelong commitment to the cause she believed in. For me, the key lesson is the importance of critics who are willing to follow her example of being so well informed that they can engage constructively in the process of change.

## References

Brambell, F.W.R. (Chairman) (1965) *Report of the Technical Committee to Enquire into the Welfare of Animals Kept Under Intensive Livestock Husbandry Systems.* HMSO, London. (The quotations are on the title page and page 2.)

Fraser, D. (2001) The 'New Perception' of animal agriculture: legless cows, featherless chickens, and a need for genuine analysis. *Journal of Animal Science* 79, 634–641.

Fraser, D. (2008) *Understanding Animal Welfare: The Science in its Cultural Context.* Wiley-Blackwell, Oxford, UK.

Fraser, D. (2012) Animal ethics and food production in the 21st century. In: Kaplan, D. (ed.) *Philosophy of Food.* University of California Press, Berkeley, California, pp. 190–213.

Harrison, R. (1964) *Animal Machines: The New Factory Farming Industry.* Vincent Stuart, London.

Thorpe, W.H. (1965) The assessment of pain and distress in animals. Appendix III. In: Brambell, F.W.R. (Chairman) *Report of the Technical Committee to Enquire into the Welfare of Animals Kept Under Intensive Livestock Husbandry Systems.* HMSO, London.

Woods, A. (2012) From cruelty to welfare: the emergence of farm animal welfare in Britain, 1964–1971. *Endeavour* 36, 14–22.

# Ruth Harrison's Later Writings and Animal Welfare Work

# 5

**Donald M. Broom**
*University of Cambridge, UK*

Ruth Harrison was my good friend during the last 25 years of her life. She was not involved in the animal production industry but she appreciated the pressures on those who were in the industry. She was not involved in politics but she knew how to present information for political decision makers. She was not a scientist but she understood the importance of science and strongly supported a scientific approach to animal welfare.

Ruth had always intended to write a second book and, at her request, her later published works and public addresses are being compiled as a book by sisters, Marlene and Diane Halverson, her American friends. I shall focus here on some prescient messages from Ruth's writings since *Animal Machines* and on her other work, in particular at the Council of Europe.

One of the key ideas put forward in *Animal Machines* (Harrison, 1964) was that some farming systems pushed animals outside their range of effective biological functioning, forcing them to try to adapt in ways that were difficult or impossible for them. Ethologist, W.H. Thorpe, like Ruth Harrison, was a member of the Brambell Committee (1965) and his explanations of animal needs fitted well with the views presented by Ruth Harrison. Some of the needs of animals are common to many species of animals, including humans. Others are specific to the kind of animal, so pigs need to root with their noses when exploring and searching for food and hens need to have a nest when they are about to lay an egg. If needs are not fulfilled, the welfare of animals is poor, as demonstrated by attempts to cope with the problems that are associated with abnormal behaviour and physiology.

© J. Harrison and J. Wilson 2013. *Animal Machines* (Ruth Harrison)

Thorpe's concept of needs (1965) was developed by Harrison (1967, 1970, 1980), referring first to the needs of mink, second to the necessity for farmed animals to be able to use their locomotor ability to walk, swim or fly and their senses, for example to see, which they could not do in dark conditions, and third to needs to show behaviour patterns that are normal for them. Animal welfare scientists also developed the concept of needs, for example Duncan and Wood-Gush (1971, 1972), Hughes and Duncan (1988) and Toates and Jensen (1991). Broom and Johnson (1993) define a need as a requirement, which is part of the basic biology of an animal, to obtain a particular resource or respond to a particular environmental or bodily stimulus. Throughout the past 25 years, the starting point for a review or for recommendations about the welfare of a species, for example by the Council of Europe or in European Union and European Food Safety Authority reports, has been a list of the needs of animals of this species, as demonstrated by scientific studies. A general guideline is provided by the Five Freedoms, but there are some problems with the concept of free-dom (Broom, 2003), so the more scientific approach is to consider needs by assessing evidence for them. Many of Ruth Harrison's later statements and publications described the results of scientific studies of the strengths of animal preferences as evidence for needs; for example, Duncan (1978, 1992), Stolba and Wood-Gush (1989), Dawkins (1983, 1990) and Matthews and Ladewig (1994). For reviews of these ideas, see Broom and Fraser (2007), Fraser (2008) and Broom (2011).

When referring to needs, Ruth Harrison often emphasized evidence from the behaviour of animals, as she was aware that the scientific estab-lishment and animal production researchers at that time frequently neglected to consider this. She sometimes used the term 'behavioural needs' but, as emphasized by Dawkins (1983) and Broom and Johnson (1993), this was not strictly correct, as the need itself was a construct in the brain so could not be called behavioural or physiological. The scien-tifically accurate, if less elegant, phraseology is 'the needs to show cer-tain behaviours' or 'the need fulfilled by a certain physiological change'. At the Council of Europe Committee (the Standing Committee of the European Convention for the Protection of Animals Kept for Farming Purposes, abbreviated as T-AP), which has produced recommendations on the welfare of farmed animals, Ruth Harrison, together with Ingvar Ekesbo, Andreas Steiger and myself, strongly supported the inclusion of details about the biology of the animal species that was the subject of the recommendation, as this provided information about the 'biological needs' of the animals.

Ruth Harrison's contributions to the Council of Europe Committee, which she attended on behalf of the Eurogroup for Animal Welfare, typi-cally involved reference to recent scientific papers which she had read. I attended this committee for 13 years as a scientific advisor on behalf of the International Society for Applied Ethology, which is the major scientific society for animal welfare scientists. Ruth appreciated at an early stage that scientific evidence was much more difficult to refute, for those defending an industry practice, than was an expert opinion. Hence, she read key papers herself and was keen to hear about evidence from scientific experts. In 1969, she said in a paper presented to the Royal Society of Health that the fact that stress was often not recognized was leading to acceptance of practices that were harmful to animals (Harrison, 1969a). She supported very strongly the development of animal welfare science from its early stages in the 1980s to the substantial discipline that it has later become.

From an early stage, Ruth Harrison emphasized the positive aspects of animal welfare. In an article in *The Observer* in 1969, she said that our farm animals were complex social animals (Harrison, 1969b). She advocated keeping them in conditions that would result in good welfare rather than making them zombies or parodies of themselves. In a paper given at the Agricultural Development and Advisory Service Conference in 1988, she said that one of the biggest curbs on progress in improving animal welfare was the attitude that the worst welfare should be prevented rather than that the best welfare should be promoted (Harrison, 1988). She commented that animal welfare science should be focused more on measuring positive welfare rather than just evaluating the worst welfare. The scientific discipline has eventually come to develop in this way.

Frustration at the slowness of progress towards good welfare in farm animals was often evident in Ruth's writings. She considered (Harrison, 1978) that voluntary codes led to much less improvement than laws. In her Hume Memorial Lecture in 1987, she stated that the recommendations of the Brambell Report had been only weakly implemented by government. In many cases, action had not been taken, with the excuse that more evidence was needed (Harrison, 1987). For example, she said that focus on the Five Freedoms, which were really just a general guideline, had led to a delay in action, with governments requesting that more scientific research be done (Harrison, 1993).

Illogical ethical positions are adopted by some with vested interests. Some of those who use animals on farms will use systems and procedures which result in welfare that is poor, to a degree that they would condemn strongly if the animal were their pet. In a lecture presented in Trafalgar Square, London, in 1965, Ruth Harrison said, 'People's attitudes to animals tend to be governed in part by the use to which they are put rather than by the animals simply as animals' (Harrison, 1965).

However, she was pragmatic and thought of what producers of farm animals had to do to survive financially; for example, saying in the same speech, 'At a time of over-production in practically all animal products in the Community, there is time to re-assess systems in favour of those that respect animal life and dignity and to direct resources in helping producers to use them.'

A frequent argument used against proposals to improve animal welfare, where some extra cost would be involved, was that food production should be maximized as there were many people starving in the world. In a paper given in Ottawa (R. Harrison, Ottawa, 1978, unpublished document), she said that 'meat production in most of the world has nothing to do with the starving poor'.

A major improvement in farm animal welfare in recent years was brought about by consumers who had learned about the poor welfare of animals used to produce specific animal products and had refused to buy those products. The result was, first, that high welfare products were advertised and, second, that retail food companies set up standards for animal welfare. This direction of change was anticipated by Harrison (1971), who stated that 'It is only by labelling that the public can make known its preferences.' The move towards sustainable agriculture (Broom, 2010) was also anticipated by Ruth Harrison's Farm Animal Care Trust, by sponsoring a symposium on the subject (Marshall, 1992).

Harrison's work has had long-standing influence and continues to be referenced today by scientists and lay people with an interest in the lives of farmed animals. The progress that has been made, especially in Europe, between 1972 and the present with respect to formulating rules and scientific rationales for farm animal protection would not have been made without her initial stimulus and subsequent influential contributions. Ultimately, Harrison's work led to a new outlook on the use of animals in agriculture and the development of animal welfare science. Yet, the reader of this volume will be impressed not only by Ruth's summation in 1964 of the challenges presented to animals but also by her prescience concerning future developments and by the amount of work still to be done.

## References

Broom, D.M. (2003) *The Evolution of Morality and Religion*. Cambridge University Press, Cambridge, UK, pp. 259.

Broom, D.M. (2010) Animal welfare: an aspect of care, sustainability, and food quality required by the public. *Journal of Veterinary Medical Education* 37, 83–88.

Broom, D.M. (2011) A history of animal welfare science. *Acta Biotheoretica* 59, 121–137.

Broom, D.M. and Fraser, A.F. (2007) *Domestic Animal Behaviour and Welfare*, 4th edn. CAB International, Wallingford, UK, pp. 438.

Broom, D.M. and Johnson, K.G. (1993) (reprinted with corrections, 2000) *Stress and Animal Welfare*. Kluwer, Dordrecht, Netherlands.

Dawkins, M. (1983) Battery hens name their price: consumer demand theory and the measurement of animal needs. *Animal Behaviour* 31, 1195–1205.

Dawkins, M.S. (1990) From an animal's point of view: motivation, fitness and animal welfare. *Behavioral and Brain Sciences* 13, 1–31.

Duncan, I.J.H. (1978) The interpretation of preference tests in animal behaviour. *Applied Animal Ethology* 4, 197–200.

Duncan, I.J.H. (1992) Measuring preferences and the strength of preferences. *Poultry Science* 71, 658–663.

Duncan, I.J.H. and Wood-Gush, D.G.M. (1971) Frustration and aggression in the domestic fowl. *Animal Behaviour* 19, 500–504.

Duncan, I.J.H. and Wood-Gush, D.G.M. (1972) Thwarting of feeding behaviour in the domestic fowl. *Animal Behaviour* 20, 444–451.

Fraser, D. (2008) *Understanding Animal Welfare: The Science in its Cultural Context.* Wiley Blackwell, Chichester, UK.

Harrison, R. (1964) *Animal Machines.* Vincent Stuart, London.

Harrison, R. (1965) (unpublished) Mass rally in Trafalgar Square against factory farming. Reported in *The Observer*, 25 April 1965, p. 4.

Harrison, R. (1967) What price mink? Unpublished pamphlet.

Harrison, R. (1969a) Intensive livestock farming and health (c) Welfare. *Proceedings of Health Congress, Eastbourne, 28 April to 2 May 1969.* Royal Society of Health, London.

Harrison, R. (1969b) Why animals need freedom to move. *The Observer*, 12 October 1969, p. 7.

Harrison, R. (1970) Unpublished discussion: Proceedings of Factory Farming Conference.

Harrison, R. (1978) Intensive livestock systems: where do we draw the line? Unpublished conference proceedings.

Harrison, R. (1980) Animal production and welfare – practical considerations. *Animal Regulation Studies* 2, 215–221.

Harrison, R. (1987) Farm animal welfare: What, if any, progress? Hume Memorial Lecture, Royal Society of Medicine, 26 November 1987. Universities Federation for Animal Welfare, Potters Bar, Herts, UK.

Harrison, R. (1988) Livestock production methods at a crossroads. Unpublished paper presented at the Agricultural Development and Advisory Service Conference, 6 April 1988, Harper Adams College, Newport, UK.

Harrison, R. (1993) Since *Animal Machines. Journal of Agricultural and Environmental Ethics* 6 Suppl. 1, 4–14.

Hughes, B.O. and Duncan, I.J.H. (1988) Behavioural needs: can they be explained in terms of motivational models? *Applied Animal Behaviour Science* 20, 352–355.

Marshall, B.J. (ed.) (1992) *Sustainable Livestock Farming into the 21st Century.* Centre for Agricultural Strategy, University of Reading, Reading, UK.

Matthews, L.R. and Ladewig, J. (1994) Environmental requirements of pigs measured by behavioural demand functions. *Animal Behaviour* 47, 713–719.

Stolba, A. and Wood-Gush, D.G.M. (1989) The behaviour of pigs in a semi-natural environment. *Animal Production* 48, 419–425.

Thorpe, W.H. (1965) The assessment of pain and distress in animals. Appendix III. In: Brambell, F.W.R. (Chairman) *Report of the Technical Committee to Enquire into the Welfare of Animals Kept under Intensive Husbandry Conditions.* HMSO, London.

Toates, F. and Jensen, P. (1991) Ethological and psychological models of motivation: towards a synthesis. In: Meyer, J.A. and Wilson, S. (eds) *Farm Animals to Animats.* MIT Press, Cambridge, Massachusetts, pp. 194–205.

# RUTH HARRISON

# Animal machines

*foreword by*

# Rachel Carson

# ANIMAL MACHINES

## The New Factory Farming Industry

FSC
www.fsc.org
MIX
Paper from
responsible sources
FSC® C013604

# Animal Machines

## The New Factory Farming Industry

**Ruth Harrison**

*Foreword by*
**Rachel Carson**

www.cabi.org

# Foreword

The modern world worships the gods of speed and quantity, and of the quick and easy profit, and out of this idolatry monstrous evils have arisen. Yet the evils go long unrecognised. Even those who create them manage by some devious rationalising to blind themselves to the harm they have done society. As for the general public, the vast majority rest secure in a childlike faith that 'someone' is looking after things – a faith unbroken until some public-spirited person, with patient scholarship and steadfast courage, presents facts that can no longer be ignored.

This is what Ruth Harrison has done. Her theme affects practically every citizen, for it deals with the new methods of rearing animals destined to become human food. It is a story that ought to shock the complacency out of any reader.

Modern animal husbandry has been swept by a passion for 'intensivism'; on this tide everything that resembles the methods of an earlier day has been carried away. Gone are the pastoral scenes in which animals wandered through green fields or flocks of chickens scratched contentedly for their food. In their place are factorylike buildings in which animals live out their wretched existences without ever feeling the earth beneath their feet, without knowing sunlight, or experiencing the simple pleasures of grazing for natural food – indeed, so confined or so intolerably crowded that movement of any kind is scarcely possible.

As a biologist whose special interests lie in the field of ecology, or the relation between living things and their environment, I find it inconceivable that healthy animals can be produced under the artificial and damaging conditions that prevail in these modern factorylike installations, where animals are grown and turned out like so many inanimate objects. The crowding of broiler chickens, the revolting insanitary conditions in the piggeries, the lifelong confinement of laying hens in tiny cages are samples of the conditions Ruth Harrison describes. As she makes abundantly clear, this artificial environment is not a healthy one. Diseases sweep through these establishments, which indeed are kept going only by the continuous administration of antibiotics. Disease organisms then become resistant to the antibiotics. Veal calves, purposely kept in a state of induced anaemia so their white flesh will satisfy the supposed desires of the gourmet, sometimes drop dead when taken out of their imprisoning crates.

The question then arises: how can animals produced under such conditions be safe or acceptable human food? Ruth Harrison quotes expert opinion and cites impressive evidence that they are not. Although the quantity of production is up, quality is down, a fact recognised in a most significant way by some of the producers themselves, who, for example, are more likely to keep a few chickens in the back yard for their own tables than to eat the products of the broiler establishments. The menace to human consumers from the drugs, hormones, and pesticides used to keep this whole fantastic operation somehow going is a matter never properly explored.

The final argument against the intensivism now practised in this branch of agriculture is a humanitarian one. I am glad to see Ruth Harrison raises the question of how far man has a moral right to go in his domination of other life. Has he the right, as in these examples, to reduce life to a bare existence that is scarcely life at all? Has he the further right to terminate these wretched lives by means that are wantonly cruel? My own answer is an unqualified no. It is my belief that man will never be at peace with his own kind until he has recognised the Schweitzerian ethic that embraces decent consideration for all living creatures – a true reverence for life.

Although Ruth Harrison's book describes in detail only the conditions prevailing in Great Britain, it deserves to be widely read also in those European countries where these methods are practised, and in the United States where some of them arose. Wherever it is read it will certainly provoke feelings of dismay, revulsion, and outrage. I hope it will spark a consumers' revolt of such proportions that this vast new agricultural industry will be forced to mend its ways.

**Rachel Carson**

# Acknowledgements

During my study of intensive rearing I have visited farms and agricultural establishments all over the country and I would like to thank the many farmers and agricultural advisers, many of whom were ardent protagonists for intensive rearing methods, for the help they gave me.

Various departments of the Ministry of Agriculture have been most patient and helpful in verifying facts for me, and I have obtained friendly assistance from the libraries of the Soil Association, the Royal College of Veterinary Surgeons, the *Farmer and Stockbreeder*, and the *Farmer's Weekly*. I have for some years been a regular and interested subscriber to *Agriculture, Poultry World* and other agricultural journals, and I have gratefully and freely quoted from their columns in support of my thesis. In no instance, however, do I imply that because the quotations come from these journals they represent the views and opinions held by the owners and editors.

I owe a great debt of gratitude to Gwen Barter who made available to me a large amount of material, and to Sydney Jennings who, in an already overcrowded life, yet found time patiently to comment on the book as it was written. His encouragement has been an inspiration to me.

I am indebted also to Dr Westlake, Dr Franklin Bicknell, Dr Frank Wokes, Mrs Mary Sharp, Frank Blackaby, Laurence Easterbrook and Mrs Christine Stevens, for their helpful comment.

My thanks are due to Dr Milton for permission to quote from correspondence, to Rachel Carson and Dr Bicknell for permission to quote from their books, and to the following publishers and editors: Hamish Hamilton, Faber and Faber, The Devin-Adair Company, Bailliere Tindall and Cox, The American Academy of Applied Nutrition, The Soil Association and *The Veterinary Record*.

Finally, I would like once more to thank my family for their forbearance, and especially my husband Dex whose help has been untiring and unfailing.

'It's GRASS y'fools – you're supposed to EAT it – Remember?'
By kind permission of *Farmer & Stockbreeder*

# Introduction

I am going to discuss a new type of farming, of production line methods applied to the rearing of animals, of animals living out their lives in darkness and immobility without a sight of the sun, of a generation of men who see in the animal they rear only its conversion factor into human food.

What is factory farming?

What is meant by intensive rearing?

Here is an explanation in the words of one of its experts, Dr Preston of the Rowett Research Institute:

> Rapid turnover, high-density stocking, a high degree of mechanisation, a low labour requirement, and efficient conversion of food into saleable products, were the five essentials for a system of animal production to be called intensive. (*Farmer and Stockbreeder*, 19th December 1961)

In other words farm animals are being taken off the fields and the old lichen covered barns are being replaced by gawky, industrial type buildings into which the animals are put, immobilised through density of stocking and often automatically fed and watered. Mechanical cleaning reduces still further the time the stockman has to spend with them, and the sense of unity with his stock which characterises the traditional farmer is condemned as being uneconomic and sentimental. Life in the factory farm revolves entirely round profits, and animals are assessed purely for their ability to convert food into flesh, or 'saleable products'.

Let me tell you about a visit to one of the more extreme units where veal calves were reared. We came out of the bright sunlight into the dark,

© J. Harrison and J. Wilson 2013. *Animal Machines* (Ruth Harrison)

windowless shed. The farmer switched on the light and there was instant pandemonium within a row of narrow, enclosed crates at one end of the shed. When the noise subsided he carefully let down the shutter in front of one of the crates and revealed a calf standing in a space barely large enough to hold it, its eyes wide and staring, its face a picture of misery. Twice a day it saw electric light when it was fed. Otherwise it dragged out its existence in the dark, cramped and motionless, barely living before it was slaughtered.

As I looked at that calf I wondered how many people knew anything of these new farming methods, I wondered how necessary they were and what possible justification there could be for them. I wondered whether there could be any goodness in the flesh as an end product. I wondered most of all that this unbelievable thing could be happening in the middle of the twentieth century when man was discovering space and so many exciting and wonderful new worlds were opening up to him. It seemed incongruously primitive for a people who prided themselves on being civilised and, moreover, on being a 'nation of animal lovers'.

Most people, especially in towns, tend to be ignorant of the processes by which food reaches their table, or if not ignorant they find it more comfortable to forget. Farm produce is still associated with mental pictures of animals browsing in fields and hedgerows, of cows waiting patiently in picturesque farmyards for the milking, of hens having a last forage before going to roost or sheep being rounded up by zealous dogs, and all the family atmosphere embracing the traditional farmyard. This association of ideas is cleverly kept alive by the giants of the advertising world who realise that the public still associates quality with healthy surroundings. A picture of the close-tethered veal calf standing uncomfortably on slats in its gloomy crate, the battery hen cramped in its cage, the closely packed, inert mass of pigs on the floor of the sweat-box piggery, or the sea of broilers in their dim shed, would not, they rightly surmise, help to sell their products.

The sad thing is that the further people get absorbed into large commercial combines and turn their minds purely to efficiency and material progress, the more they must sink their consciences or salve them with woolly thinking. You will be able to see how this is happening in the field of factory farming. Life is cheap to the factory farmer. He is dealing with so many animals that 'culls' are an everyday occurrence. To cull is to take out an animal which is not making a profit and to kill it. The principle of exterminating less hardy animals starts right from the beginning. Thus a hatchery is advised to cull all chicks which hatch somewhat later than the others, not because they are deformed in any way, but because they are stragglers and therefore assumed to be weaklings. Chickens can be turned out in their millions and are therefore considered a more expendable commodity than larger and more expensive forms of livestock. But the principle is the same, the cheapness of life itself, the lowering of standards.

A danger of accepting any form of life as cheap is that each successive generation might accept slightly lower standards.

How far have we the right to take our domination of the animal world? Have we the right to rob them of all pleasure in life simply to make more money more quickly out of their carcasses? Have we the right to treat living creatures solely as food converting machines? At what point do we acknowledge cruelty?

To some extent, as the Minister of Agriculture is so fond of telling us, farm animals have always been exploited by man in that he rears them specifically for food. But until recently they were individuals, allowed their birthright of green fields, sunlight and fresh air; they were allowed to forage, to exercise, to watch the world go by, in fact to live. Even at its worst, with insufficient protection against inclement weather or poor supplementation of natural food, the animal had some enjoyment in life before it died. Today the exploitation has been taken to a degree which involves not only the elimination of all enjoyment, the frustration of almost all natural instincts, but its replacement with acute discomfort, boredom and the actual denial of health. It has been taken to a degree where the animal is not allowed to live before it dies.

For the factory farmer and the agri-industrial world behind him, cruelty is acknowledged only where profitability ceases. If an animal continues to grow and put on flesh, even when this depends on the heavy use of drugs, he deems that his treatment of it cannot be said to be cruel, though it be crated up in the dark all its life. The law relating to animals is loose and ill-defined and, moreover, hopelessly out of date. So we fall back on public conscience, easily roused and as easily lulled. Whether it be ignorance of these methods or just lack of thought on the part of the public, factory farming has paid off as far as the industry is concerned and each year sees the introduction of new niceties and the exploitation of ever more animals.

What do veterinary surgeons think of all this? Are they willing and able to keep a check on the industry? Do they have an oath like the Hippocratic oath compelling them to help all animals, or have they become merely the administrators of yet more drugs?

> ... The present scale of veterinary fees is so miserable that no vet can afford to antagonise his farmer clients by accusing them of cruelty every time he visits. I personally am glad to have an entirely small-animal practice, with all its quota of false sentimentality, than an agricultural one,

wrote a veterinary surgeon in *The Observer* (6th May 1962).

The factory farmer cannot rely, as did his forebears, on generations of experience gained from the animals themselves and handed down from father to son; he relies on a vast array of back-room boys with computing machines working to discover the breeds, feeds and environment most suited to convert food into flesh at the greatest possible speed, and every batch of animals reaching market is a sequel to another experiment or part of an experiment. This type of research, and the training given on these lines in agricultural schools, gives rise to some misgivings as to the future of animal husbandry. Does the agricultural student, even with the support of allied industries, know enough about the physiology of the animal and is his training wide enough?

It is of interest to read the following letter from D. H. Armstrong, M.R.C.V.S., in *The Veterinary Record*, 23rd December 1961:

> Early this year Miller (*Vet. Rec.* 73, 2) commented on the superficial training on the animal side, given in the courses in agriculture available in Britain. In the circumstances it is surprising, if not ridiculous, to find that almost all activities having to do with animals in agricultural teaching, advisory and research projects (excepting veterinary education and practice) are in the hands of holders of one or more of the very varied agricultural qualifications.
>
> A good example came to hand recently in the second Report of the Hill Farming Research Organisation. On a Board of Management of seventeen there is one veterinary surgeon, but a staff of twelve listed as engaged on animal studies lacks even this meagre representation. …
>
> On page 10 the Report states that the organisation is not equipped to undertake critical studies in animal health; but how it is possible to divorce these from animal production is a secret known only to the non-veterinary animal husbandry expert.…
>
> At a time when so much lip service is being paid to the value of specialised science training it is disturbing to find administrative departments so ill informed or poorly advised as those responsible for work on farm animals appear to be … it must be disquieting to note that the animal-scientist of the public and farming press, the animal nutrition advisor of industry, and the animal husbandry specialist and research worker in many institutions, frequently lacks the training in anatomy and physiology that we believe to be essential for the understanding of the results of any serious work on animals.…

It is obvious to anyone, let alone a veterinarian, that the conditions under which intensively kept animals are reared indoors could not possibly lead to healthy animals, and animal health has deteriorated to a degree such as to cause acute anxiety to the veterinary world. But just as people today tend to regard doctors as miracle men able to overcome with drugs the effects of any foolish living, so these farmers regard veterinary surgeons as the miracle men of their world able to produce drugs to combat the effects of flouting every natural need of the animal. If the drugs happen to make serious physiological changes this type of farmer does not concern himself as long as the carcase remains saleable.

To keep animals alive in the conditions in which they are reared antibiotics are incorporated in their feed, heavier doses of drugs given at the least sign of flagging, and growth stimulants, hormones and tranquillisers all have their part to play in the forcing of rapid conversion of animal feeding-stuffs into flesh.

> The roast beef of old England may have come from a stall-fed creature that is given oestrogens and aureomycin during its period of growth, then received tranquillisers to ensure a docile journey to slaughter, just before which it is given intravenous antibiotics so that the meat will keep without cold storage and become tenderer during this period. Residues of all these may be in the meat.

So stated Dr Hugh Sinclair, M.A., D.M., M.R.C.P., the nutritionist, at a talk given to the Second Conference on the Health of Executives, 1960.

Other governments have banned the use of hormones and other chemicals in the raising of food as potentially dangerous to the consumer. Our own government does not think there is any danger.

All the time, as I have shown later in the book, the incidence of degenerative diseases in man rises at an alarming pace. Now I am not suggesting that this increase is attributable wholly to factory farming methods, far from it, but I do suggest that there is a contributory hazard in the heavy, continuous and universal use of drugs in these systems through residues in the meat, and I cannot help wondering how thoroughly all the drugs are tested for their effect on man before being used. For example, the antibiotic chloramphenicol has been tried out as a growth stimulant for calves. It was found to be less successful than chlortetracycline and oxytetracycline, but that does not give us any assurance that it is not being used any more. The Association of Clinical Pathologists announced on 16th March 1963 (reported in *Sunday Times* on 17th March) that '27 persons in this country have died from diseases of the blood directly attributed to the antibiotic called chloramphenicol … its use is justified only in the treatment of life-endangering infections when no other effective antibiotic is available'. I have often also wondered whether there may be a possibility of disease being transmitted direct to the consumer from the carcases of these unhealthy animals. Leukaemia and leucosis, for example, are related forms of cancer. Could there be any connection between the rise of leukaemia in man and the 'explosive outbreaks' of leucosis said to exist in broiler houses?

A comforting theory that has often been put forward is that the general standard of nutrition is high in this country so that some inferior or even slightly deleterious food does not present a hazard to the average consumer. But many far-seeing scientists have for years been trying to show us that not only is there no safety margin but in fact a very real danger in the way our food is produced.

Leonard Wickenden, Fellow of the American Institute of Chemists, writes in his book *Our Daily Poison*:

> It almost looks as if we are risking our lives in vain, and the use of the words 'risking our lives' is no indulgence in hyperbole. If few of us are meeting sudden death through the daily absorption of small quantities of

a variety of poisons, many of us are falling victims to serious diseases and there are probably few of us whose health is not, to some degree, affected.

Meat eating has become a hazard and this is only because of the urge for quick profits.

We would seem to be placing a very heavy burden of responsibility on the farmer and the feed manufacturer; their chief concern is with rapid feed conversion but how much do they really know of the ultimate effect on the human consumer of all the drugs they use? Yet, for example, feeding firms are entitled to incorporate up to 100 grammes of antibiotic in each ton of animal feeding stuffs as a regular supplement for intensively kept animals, and the farmer may buy and incorporate antibiotics at any level he thinks fit. This, when the merest trace can mean the difference between life and death to a consumer.

The September 1962 issue of *Agriculture* said that 'in the U.S.A. 87 per cent of agribusiness is an off-the-farm operation' and warned the producer that he must take an active interest in all stages of the production-processing-selling chain. This chain has made the agricultural industry the largest in this country also, and one in which the power of vested interests and the fear of losing livelihoods makes silent many who are otherwise seriously disturbed by some current farming practices.

The whole structure of our society is so bogged down with vested interests that to get a true picture of what is happening in any particular field is extremely difficult. I have found this even in my small field of research. I have met with evasion after evasion and have had to build up my knowledge and piece together the truth bit by bit as I discovered it, marvelling sometimes at the distortion of fact which I was supposed to be swallowing.

'We are only producing what the public wants.'

This brief quotation seems to sum up the reasons which impelled me to write this book. I was originally drawn to the subject because I felt the modern methods by which farm animals are increasingly being reared are, quite frankly, cruel. But the more I studied the subject the deeper became my conviction that other issues are involved. The degradation of the animal in the appalling ways it is now made to eke out its existence must have an impact on human self-respect, and ultimately on man's treatment of man – 'Inasmuch as ye have done it unto one of the least of these my brethren ye have done it unto Me.' Some find it easy to lull their consciences when only animals are concerned, but the issues under discussion extend beyond conscience and impinge in the most practical manner on the physical well-being of the human race in so far as the food produced by these means is not only inferior but dangerous.

# Broiler Chickens

The poultry industry, for such it has become, has grown out of all recognition in the last twenty years, and especially rapidly since 1955. Before the war holdings were small, and with the war came rationing of feeding stuffs and consequent limiting of flock numbers. The end of rationing, however, resulted in the growth of very large flocks and the expansion of both the egg and the table poultry industry. The expansion of broilers has snow-balled very rapidly. In 1954 there were 20 million broilers reared, in 1957 56 million, in 1959 108 million, in 1960 142 million, and it has been hope-fully estimated in the industry that by 1965 they should be able to produce upwards of 200 million broilers a year. These have now largely replaced the former capon trade, which still exists, however, on a much reduced scale.

> Production in the U.S.A. is expected to top two billion mark this year and houses are still going up (reported Mr W. G. R. Weeks of Durham University economics department on a visit to the States (*Farmer and Stockbreeder*, 16th May 1961)). Large-scale broiler producers are just waiting for our import restrictions to be lifted and they will bury us in chicken. One organisation in Georgia claims they have *averaged* a conversion ratio of 2·35 over 55 *million* broilers processed last year. That concern alone reckons it could feed not only West Germany, Switzerland, Hong Kong, but us as well, and still have enough over to put a chicken in every American's pot!

A billion, American style, is a thousand million, but at that it repre-sents a lot of chickens.

Hens, chickens, capons, poultry, and now broilers! This ugly example of potted jargon has been coined to describe an intensively reared chicken brought to the table when still quite young and small. I suppose it to have been derived from the verb 'to broil', meaning to roast on a spit. For a long time many people confused the term with 'boiler', the boiling fowls – those wornout, aged birds.

Whereas before the war the farmer bought in feeding stuffs, reared most of his own replacements and relied on his own proved market for his birds, the industry today is so huge that the individual farmer cannot operate without a link with other sections of the industry. So the hatchery gets its eggs from its own breeders, incubates them and sells the day-old chicks to contracted farms, which in turn rear the chicks for some $9\frac{1}{2}$ weeks and then sell them to packing stations, which in turn sell their stock to large retailers, mostly supermarkets. All are so interdependent that it has been found to be more profitable to have large combines, or 'integration' linked financially as well as by necessity, and so we have hatchery – broiler – packing-station combines, or broiler – feeding-firm – packing-station combines, or supermarket – packing-station – broiler combines; the variations can be many but the idea is the same – there must be a vast financial tie-up for the industry to be able to surmount the vicissitudes of its trading. This example of a 'take-over', reported in *Farming Express* (30th August 1962), will illustrate the complexity of these tie-ups:

> The Ross Group, with its latest take-over of Fairbairn-Chunky for £2,200,000, has now become the largest chicken producer in Europe. It has spent more than £6,000,000 in 16 months to buy up big established breeding firms.
>
> … The group's annual turnover in poultry would be between £15,000,000 and £20,000,000.
>
> It would have a weekly output of 2,000,000 chicks.
>
> Twenty-five per cent of all eggs eaten in Britain would be produced from Ross laying-stock.
>
> The group would handle 500,000 broiler chickens every week at its own processing plants.
>
> Eighty per cent of the broiler chickens eaten in Britain would come from broiler breeding stock supplied by Ross.
>
> Said Mr Ross: 'The acquisition of Fairbairn-Chunky will develop to a fine degree Ross Group integration in the poultry industry.
>
> 'Duplication of research, breeding and distribution, will be avoided, and in this way it will help to reduce the prices of chickens and eggs to the housewife.'
>
> The take-over adds Fairbairn-Chunky breeding farms in Cumberland and Scotland to those of Sterling Poultry Products, bought last year for £3,700,000.
>
> But in February, Sterling had bought up Spinks of Easingwold, a 50-year-old family firm of breeders. This brought into the Sterling group four breeding and research farms.
>
> Chunky Chicks – which has international links with two American firms, Heisdorf and Nelson, of Seattle, and Nichols Inc., of Maine, and with

Lohmann and Co., of Cuxhaven, Germany – merged with E. F. Fairbairn, of Carlisle, only last February.

*The Financial Times*, 24th August 1962, reported of the take-over:

> The Ross-Fairbairn merger, announced yesterday, completes a second revolution in the British broiler industry. The first, which occupied the 'golden years' from 1953 to 1960–61, saw the growth of very large processing concerns … as sales of table-ready birds grew from nothing to well over 100m. chickens a year. The new group, however, which will now dominate the industry, is the first which has been built up deliberately to integrate the whole production process from the hatchery to the packed and branded bird.

A warning of the possible effects of over-integration, as is happening in America, was given in the *Bedfordshire Times and Standard* (16th February 1962):

> The same trend is becoming obvious in England. If it continues, before many years are past broiler production will be in the hands of a few giants, who will operate on an international scale (even more than they do now).
>
> The discovery, already made, that broilers can be produced more cheaply in countries with lower labour costs, will then be acted on, and our own broiler industry will have neatly committed suicide.

'We don't ill-treat them,' said a broiler manager, a likeable young man with a great capacity for hard work, 'they live in a nice warm atmosphere out of the wind and the rain, and have ad lib. food. Rather like a club.' Rather like a club – Shall we examine this touching simile a little more closely?

Hatcheries draw their eggs from breeding stock kept on free range or in deep litter houses. Very few breeders keep their hens in batteries because the chicks must be supremely healthy to cope with their subsequent factory conditions in broiler or battery houses. To this end also all weaklings amongst day-old chicks are destroyed by the hatchery and only 100 per cent perfect chicks passed on to intensive units. The hatcheries give a 2 per cent mortality replacement addition free to the broiler manager, i.e. when sending 10,000 chicks for a broiler house, they will include an extra 200 to replace those which inevitably cannot stay the course. Two to six per cent mortality is taken for granted. Higher mortality gives rise to alarm.

The day-old chicks are installed, eight or ten thousand at a time, sometimes more, in long, windowless houses punctuated only with extractor fans in serried rows along the ridge of the roofs, and air intake vents along the side walls. In a big establishment these sheds will be ranked side by side each with its giant feed storage hopper standing as if on guard at one end, the whole array looking like an incongruous factory, sprouting, for no apparent reason, in the middle of some remote field.

When the industry was young a converted barn or any other odd part of existing farm buildings would be utilised to house the new-fangled broilers. There would be windows and natural lighting and ventilation, but the space available, and thereby the number of chickens, would be limited. This could not last, it wasn't efficient. Better houses could be purpose-made for the job. Artificial ventilation could be introduced to take the place of natural, more chickens could be got in that way. But, with more chickens in, they begin to fight, so eliminate the windows, that will stop them, they can't see to fight each other in the dark. So from one technological improvement to another until we emerge with the latest thing in houses, all nicely controlled and automated, the most advanced in the world.

Inside a house the impression is of a long, wide, dark tunnel disappearing into the gloom, the floor covered with chickens as far as the eye can see. There are lights down each side, hoppers hang from the beams for food, and pipes keep a constant supply of water. The houses are sprayed regularly with insecticides to keep the chickens free of pests. For the first two weeks the chicks are kept under warm brooders at a steady temperature of 90°, that of a mother hen, in a constant, round the clock, bright light. Thus they are encouraged to eat and grow quickly. After two weeks the lights are changed to amber and go on and off for two hours round the clock. So the birds eat and sleep, eat and sleep, eat and sleep. At six weeks they are big enough to feel the intensity of crowding and too much light would mean too much fighting, so the lights are changed to 25 watt red and these go on and off round the clock every two hours. Even the red lights give an impression of virtual darkness and the eyes strain until they become increasingly aware of the great sea of chickens punctuated by the hoppers, now raised from the ground to allow more room. The general flurry away from the door caused by our entrance subsides and a great stillness descends on the house. So they exist for the last four weeks of their short lives, in darkness, almost immobile, their only function to put on weight.

The birds live on a deep layer of wood shavings, known as deep litter. This gets saturated with droppings and has to be turned regularly to keep it dry. The fans do something to draw off the smell of ammonia but it lingers very strongly in the houses. The very warm atmosphere does not help in eliminating this smell which is usually quite overpowering when one first enters the house from the fresh air outside.

A whole body of literature has grown up to cover the problems involved in the caring for and maintenance of broiler houses. As might be imagined, these are anything but simple. For example, a Ministry of Agriculture booklet, *The Broiler House*, warns that electric wiring must be carefully installed because:

> *As the atmosphere in the broiler house is dusty, humid, and charged with ammonia*, it is likely to cause an electrical breakdown unless suitable wiring and fittings are used. The whole of the wiring should be carried out with damp and dust-proof fittings. Where leads are provided to loose

equipment such as brooders, these should be of the rubber armoured type normally found outdoors on building sites. Earthing of all apparatus is most important. Only a competent electrical engineer should test a circuit and say if the earthing system is effective. Because of the special atmospheric conditions in a broiler house the electrical installation should be planned in the first place, and subsequently inspected at least once a year by a trained electrician. A main switch to cut off all the power in the house should be provided, and should be placed so that it can be easily reached by a person near the entrance door. *Improvisation is particularly dangerous in the conditions which are normally found in a broiler house.* (The first italics are mine.)

It is all the more important to have no electrical breakdown in a tightly packed house because of the hazard to the chickens when workmen are in the houses. It was reported in the *News Chronicle* on 4th June 1960:

> When an electrician was summoned recently to do a job in a broiler house he was nearly sick. He could not walk across it without treading on birds. But the owner said: 'Don't worry about that. Just tread on them.'

The broiler house has a carefully controlled heating system, usually under the floor. But this, as in all intensive units, is a one-way control, it keeps the temperature up to a certain level, but refrigeration is too costly so that the temperature *cannot be kept down and the animals suffer the full brunt of a heatwave.* Fortunately for them in this country that does not often become a hazard!

The birds have access to water from pipes running through the house and are fed a compounded food to appetite. This is not a misprint, it is the jargon used in the industry, on a par with 'conversion ratio' and like felicities. It means that they can have as much food as they choose. This food is bought in bulk, stored in giant hoppers outside the house, and fed through to the hanging food hoppers inside, automatically in most cases.

This then, is the broiler industry, brought over from America, and looking still to America for leadership and advice. It is an industry with its full quota of 'jobs for the boys': Mr Robinson, an Essex broiler breeder, pointed out to an Oxfordshire conference (*Farmer and Stockbreeder*, 8th August 1961):

> So many of today's specialists in the various aspects of the poultry industry never had a thing to do with the industry 18 months to two years ago. Now they have blossomed out into what big business calls executives, advisory officers, field officers, flock farm supervisors, hatchery managers and packing station managers. I have not even mentioned the tycoons themselves or the public companies which they have formed to carry their financial burdens.

Every stage in the process is backed by a large and devoted body of research involving the back-room boys of many industries, in laboratories and experimental farms, all working assiduously towards the same end: the production of the largest bird, in the cheapest way, in the shortest possible time. The Plant Report on fowl pest, 1962, explains how this can be done:

> To produce stock for broiler chicken production it has become the practice to use a cross of which the female parent is a good egg producer, to keep chick costs down, while the male parent supplies other qualities, particularly rapid growth and a relatively high proportion of meat in the carcase.

A few days knocked off the tiny life span of the broiler chicken is greeted with enthusiasm by the industry. The *Daily Mail* reported pithily (6th April 1961):

> QUICK CHICK WILL CUT DINNER BILL
>
> Cheaper chickens for British dinner tables were promised yesterday by Mr Henry Saglio, American chicken-breeding chief, whose farms in Connecticut control 70 per cent of the world's breeding stock production.
>
> He announced in London a new strain of broiler chicken, specially bred with white skin and white legs to suit British tastes. These will grow to $3\frac{1}{2}$ lb. in nine weeks, instead of the ten weeks of present strains.
>
> Reduction in the life-cycle will save fourpence a bird to the farmer, and an improved food-conversion ratio will add an additional threepence. These together will enable producers to put birds in the shops cheaper than ever.

Without the research which has gone on behind the scenes in the broiler industry, it could not have achieved the great production numbers it has achieved today. Breeding, food conversion, growth additives, drugs, environment, light patterns, all play their part and are the subject of countless tests and experiments; *Poultry World* (2nd August 1962) wrote of an experimental station:

> The testing possibilities are infinite. But the types of test carried out generally come under four main heads. They are: (1) Food tests, (2) strain tests, (3) pharmaceuticals, (4) management. And as examples under each head, the test can be:–
>
> 1. (a) of one feed against another, or several others.
>    (b) of optimum feed-change times.
>    (c) of types of feed, e.g. mash versus pellets.
> 2. (a) tests of one broiler type against another.
>    (b) tests of one strain against another.
>    (c) tests of relative performance of cockerels and pullets.
> 3. (a) the comparative testing of coccidiostats.
>    (b) the testing of one antibiotic, or other growth-promoting additive, against another.
> 4. (a) tests to determine optimum bird densities.
>    (b) tests to determine the effect of old as compared with new litter.

(c)  tests to determine the effect of de-beaking.

(d)  tests to determine optimum light intensities.

In a broiler breeding programme carried out on Essex farms (*Farming Express*, 28th September 1961) work

> … is directed by the findings of computers in the United States. It is a programme based on tens of thousands of figures. The result is a bird which, under commercial conditions, reaches a liveweight of $3\frac{1}{2}$ lb. in 63 days with a feed conversion rate of 2·3…. 70,000 chicks a year are wing-banded…. The performance of every one of these chicks, its parents and grandparents has to be recorded and tabulated. The figures are sent to America and put through the computers. The results, together with recommendations … are returned … in a one-inch thick stack of figure-filled sheets. On those recommendations and figures she (the broiler breeder) has to select the birds for the next stage in the breeding programme. For the next generation of parent stock, a ruthless high-pressure selection programme is followed.

Mr J. Clark, economist with the North of Scotland College, spoke on the economics of broiler growing.

> Feedstuffs obviously dominate the picture, although other factors are important. A very slight variation in the conversion ratio could seriously affect the farmer's pocket. A variation of 0·1 per cent in the ratio in a 10,000-bird house producing four batches per year could mean a difference of £200.
>
> … A loss of half an oz. of food per bird per week could alter the food ratio by 0·1 per cent, with the loss already indicated. (*Poultry World*, 12th October 1961)

The feed also contains a small percentage of penicillin (10 mg. per ton) to promote growth. This is pushed up as soon as any disease is suspected.

Many broilermen have their chickens de-beaked to avoid feather peck-ing and fighting in the later stages. You can see what this removal of part of the upper mandible looks like in Fig. 7. It has been suggested that complete removal of the upper and lower mandibles would be a still better idea. The chicken would have no possibility of damaging its neighbours and could still manage pellets of food. How it would drink is not described. Another idea, adopted by the Japanese, and not without a droll touch of humour, is to provide the chickens with spectacles. Tinted red, they are intended, by neutralising the colour, to prevent the bird pecking at the red combs of its fellows. Since in this country dim red lighting is used anyway for the over-all environment, the spectacles are superseded by blinkers, also attached to the beak, which prevent the bird seeing to either side.

Feather-pecking is called a 'vice' in the poultry world, and chickens, I have been told many times, are vicious creatures, predisposed to this.

Dr Konrad Lorenz, the eminent naturalist, of the Max-Planck-Institut, Seewiesen, Germany, explains to us the laws of the farmyard:

> Do animals thus know each other among themselves? They certainly do, though many learned animal psychologists have doubted the fact and indeed denied it categorically…. This can be convincingly demonstrated by the existence of an order of rank, known to animal psychologists as the 'pecking order'. Every poultry farmer knows that, even among these more stupid inhabitants of the poultry yards, there exists a very definite order, in which each bird is afraid of those above her in rank. After some few disputes, which need not necessarily lead to blows, each bird knows which of the others she has to fear and which must show respect to her. Not only physical strength, but also personal courage, energy and even self-assurance of every individual bird are decisive in the maintenance of the pecking order. This order of rank is extremely conservative. An animal proved inferior, if only morally, in a dispute, will not venture lightly to cross the path of its conqueror, provided the two animals remain in close contact with each other…. (*King Solomon's Ring*)

This is the order of the farmyard, of free range chickens, and rarely, I have been told by free range farmers, do they have serious trouble with their hens. What trouble they do have is easily seen and therefore quickly remedied, the bully can be removed. There are bullies in all the hierarchy of life, even in man.

These 'vices' belong mainly to intensive conditions and are not inherent in the birds. Let me quote from the *Farming Express*, which in its issue dated 1st February 1962 warned poultry keepers:

> Feather-pecking and cannibalism easily become serious vices among birds kept under intensive conditions. They mean lower productivity and lost profits.
> Birds become bored and peck at some outstanding part of another bird's plumage; …
> While idleness and boredom are pre-disposing causes of the vices, cramped, stuffy and overheated housing are contributory causes. And faulty nutrition is considered by some specialists to lead to cannibalism….

and *The Smallholder*, 6th January 1962, stated:

> Feather-pecking and cannibalism has increased to a formidable extent of late years, due, no doubt, to the changes in technique and the swing towards completely intensive management of laying flocks and table poultry.
> Feather-pecking itself is of little account, but it must be borne in mind that it is often the forerunner of cannibalism, and for that reason it should be regarded as a dangerous vice.
> When it occurs, immediate steps must be taken to find the cause – *generally a fault in management* – and remedy it before the more serious habit of cannibalism becomes established.
> The most common faults in management which may lead to vice are boredom, overcrowding in badly ventilated houses, too low perches, exposed

nests, poor feathering in chicks, lack of feeding space, unbalanced food or shortage of water, and heavy infestation with insect pests....

*When flocks are well housed, correctly fed, and under constant observation, vices should not occur* (The italics are my own.)

Feather-pecking and cannibalism are but minor hazards compared with others facing the broiler manager. Disease is probably the greatest. *Poultry World* (26th July 1962) reported a talk given by Mrs K. M. Smith, head of Associated Broiler Breeders Ltd., Veterinary Division, in which she stressed again the importance of good management in broiler breeding and pointed out that in the adverse conditions under which the birds were reared it was touch and go as to whether the venture could at any time be successful:

Mrs Smith emphasised the point that although the grower had to keep his birds close on the floor and feed them forcing types of food in order to get results, such factors could lead to stress unless management was of the highest order.

Although a number of new diseases had made an appearance since broiler keeping began, generally speaking the organisms which caused trouble were the same today as had been associated with poultry for many years, she said.

It has been shown, she added, that when animals were subjected to adverse conditions a chain of events was initiated in the body, irrespective of the nature of the stress and, if this continued long, the animal developed clinical signs of disease.

'We keep our birds so close together in a house that this is in itself a stress. Moreover, at the same time we are giving them as much concentrated food as they are capable of consuming.

'Under these conditions it is rather like walking on a tightrope. Unless our management is of the highest order we are likely to run into trouble but when the management is really good then we get excellent results.'

A particularly important example of stress leading to disease was C.R.D. (chronic respiratory disease). Most important factor in this connection was the correct use of ventilation.

Mrs Smith pointed out how *coli septicaemia* was caused by an organism which was the normal inhabitant of the gut. But it could lead to disease when birds were subjected to stress.

'We have got to remember we are pumping food into the birds as hard as possible and unless we have a really adequate water supply and general management is first class they will have quite a difficult job to get rid of the normal toxic waste products from the food.'

Conditions are so crowded that any disease can sweep through the house very rapidly. 'I am allowed an extra 2 per cent day-old chicks from the hatchery,' said the broiler man, 'to allow for mortality.' We noticed the pile of dead birds by the door. 'They suffer mostly from respiratory diseases and cancer,' he added, 'but are rather too young when they are sent to the packing station to be seriously affected by any disease.' They are fed a small amount of antibiotic in their feed to suppress disease.

Fowl pest is the greatest dread of a poultry farmer's life. It can sweep through these intensive units very rapidly and slaughter of the entire flock on a farm has been compulsory. Compensation was assured to all farmers whose flocks were destroyed and the compensation was payable out of the taxpayer's pocket.

Before the Second World War there were only two outbreaks of fowl pest, or Newcastle disease as it was called, after the first outbreak near Newcastle-on-Tyne. One was in 1926 and spread to eleven counties. There were virtually no survivors amongst birds contracting the disease. The second outbreak was in 1933 on a farm in Hertfordshire. The owner slaughtered the diseased flock and the outbreak was over. It was only after the war when intensive rearing was made possible by removal of rationing, that the disease became uncontrollable. C. W. Scott wrote in the *Daily Telegraph* (30th July 1962):

> ... it is the concentration of poultry into large intensive units that has really brought about today's impossible position.
>
> This concentration has had two important effects on fowl pest.
>
> In the first place it has meant that when an outbreak occurs much larger numbers of birds have to be destroyed for each outbreak. But more important still is the greatly increased risk of spreading the disease.
>
> Fowl pest virus is easily wind-borne and the modern intensive poultry house becomes a virus factory as soon as an outbreak occurs because its forced ventilation system just pumps virus out of its chimneys and around the countryside each time an outbreak occurs.

This opinion was confirmed by a Ministry of Agriculture spokesman:

> Broiler units were the biggest single risk of the disease spreading.... So many poultry were in a confined space, and extractor fans in broiler houses carried the virus out and into the wind, causing a great risk to all poultry in the vicinity. (*Wiltshire Gazette and Herald*, 8th March 1962)

In the year 1954/55 there were 52 million hens kept and 20 million broilers reared. Amongst these there were 550 outbreaks of fowl pest, 7,000 diseased birds were slaughtered and 328,000 contacts, but otherwise apparently healthy birds, were slaughtered. By 1960/61 of the 63 million adult hens kept and the 142 million broiler chickens reared, 57,000 diseased birds were slaughtered and 1,937,000 hens and 3,013,000 broilers which were healthy but contacts were slaughtered. This meant that compensation paid rose from £364,000 in 1954/55 to £3,391,000 in 1960/61. Compensation was only payable on *unaffected* birds slaughtered. The sooner therefore that the farmer reported the disease on his farm, the less chance the disease had of spreading and the smaller his own loss would be, because whereas the whole flock was destroyed, he could recoup compensation for those of his birds which did not actually suffer from the disease and those amounted to 99 per cent of birds slaughtered. By 1961/62 the disease had got really out of hand with a compensation bill facing the taxpayer of £5·3 million for the first six months.

Another reason for this steep rise was suggested by a farmer in Norfolk, one of the districts where most compensation was paid out. He told a *Daily Telegraph* reporter (1st November 1961):

> 'Fowl pest has broken out time and time again on the same farms and often the number of birds destroyed was larger than on the previous occasion. It has become common knowledge that some breeders have come upon fowl pest as an easy way to make a fortune....
>
> They have reared far more birds than they could possibly hope to sell on the open market, encouraged overcrowding, and then sat back waiting for fowl pest to be confirmed. As a result they have got the market value of their birds without the trouble of marketing them.' ... He had been inundated with telephone calls.
>
> 'Some have been from people who told me to mind my own business, but most have been from farmers who also think it is time that this great scandal was brought into the open.'

Another farmer from Norfolk was among the latter.

> 'Compensation has been exploited on a fantastically large scale and the taxpayer has had to pay the bill.' ...
>
> A Ministry of Agriculture spokesman said last night: 'It is true that we have been worried about certain aspects of fowl pest. There have been cases recently where compensation to breeders has been withheld pending full investigations.'

Disquiet was not only felt by some farmers; a lively discussion took place in the House on 20th March 1962 in which some Members made it clear that thoughts of the kind expressed by the Norfolk farmers had also worried them. I quote extracts from the Parliamentary Report in *The Times* of 21st March:

> MR HAYMAN (Falmouth and Camborne, Lab.) asked whether the Ministry were justified in continuing to pour out these prodigious sums in the way they were doing. The Minister had told the House on an earlier occasion that one man got a third of a million pounds in compensation. He hoped that the Minister would say that a thorough investigation had been made, and let the House know its results.
>
> Year after year there were huge claims for compensation for fowl pest. It seemed to him that for this kind of thing to be going on year after year was simply the result of lax administration.
>
> MR BULLARD (King's Lynn, C.) said that Mr Hayman seemed to be alleging fraud.
>
> MR HAYMAN – It seems to me there may well be fraud. I do not suggest there is. There seems ground for suspicion.

Mr Bullard said a warning ought to be given to poultry keepers that if they massed birds in these huge flocks, and there was an outbreak of fowl pest, they ran the risk of not being paid compensation for them. He did not see why poultry keepers should be compensated for flying in the face of one of the elementary risks of over-intensification.

Mr Kenyon (Chorley, Lab.) said that when a farmer could buy poultry in the market to restock his farm and get more in compensation than the amount he had to pay there was a tremendous temptation. The assessors of poultry did their best, but in that industry there were some very smart dealers. He hoped the regulations would be made a little more stringent.

However, the Minister of Agriculture and the Secretary of State for Scotland had already, in 1960, appointed a Committee under the Chairmanship of Professor Sir Arnold Plant to look into the whole question of fowl pest and how best to tackle it, and in 1962 they produced their Report.

This Report surveyed the history of the disease both in this country and abroad, examined methods used so far in its eradication, and came to the conclusion that:

> ... *it would continue to be appropriate for the taxpayer to meet the cost of the veterinary service and of destroying and disposing of affected flocks, but we recommend that in future the cost of compensation should be met by the poultry industry.* This arrangement would provide an important additional incentive to a general use of vaccine, since producers will be aware of the importance of reducing to a minimum, in their own interest, the need to destroy large and valuable flocks of apparently healthy birds if the cost of compensation is to be held within a tolerable limit.

The gist of their recommendations was as follows:

> We have concluded that control of the disease rather than eradication should be the immediate aim....
>
> We are satisfied that a ... poultry industry could be maintained ... if loss from Newcastle disease were limited only by the voluntary use of suitable vaccines....
>
> ... The attempt should be made to combine the continuance of slaughter to contain the disease, with vaccination to limit the losses from it.
>
> ... And the cost of the vaccination should be met by the poultry producer.
>
> ... The taxpayer to meet the cost of destroying and disposing of affected flocks ... but the cost of compensation should be met by the poultry industry.

It was estimated that the cost per bird of vaccination subsidised by the Government would be $\frac{1}{2}$d. to the producer, whereas on 1960 figures the cost of compensation, if spread over every type of chicken, worked out at $4\frac{1}{2}$d. per bird, paid by the Chancellor out of the taxpayer's pocket.

A report in *Poultry World* (14th February 1963) told of the Egg Board's concern that with cessation of compulsory slaughter only two months away, less than twenty per cent of poultry keepers had vaccinated their flocks. It emphasised the grave risk producers were taking and the severe financial loss which could result to them from not protecting their flocks. A Lancashire branch of the N.F.U. pointed out that two lots of vaccinations (for laying birds at three weeks and point of lay) were not enough, and recommended a third dose at nine or ten weeks.

A possible reason for lack of enthusiasm to vaccinate was given in a letter in *Poultry World* (14th February 1963):

> Our chairborne administrators seem to be quite oblivious of the practical difficulties and expense which the smaller poultry farmer faces in vaccinating his flock against fowl pest.
>
> The majority of small producers employ no labour and it is therefore almost essential for them to hire extra help to assist with the actual vaccination. But the farmer himself will more often than not have to catch and crate the flock single handed the previous night, as catching in daylight is not practical.
>
> If anyone is unaware what this involves let him try catching and crating 500 to 1,000 birds single handed from upwards of six houses by torch light. Even with two people it is a formidable task and a further difficulty is that few small farmers possess sufficient crates to hold all their birds.
>
> When it is also apparent that the vaccination has to be undertaken more than once to be really effective it should surprise no one that there has been so little response from the smaller farmer. The cost of the vaccine is a mere fraction of the cost and effort involved in carrying out the work.
>
> In view of the practical difficulties involved, one cannot help wondering whether it would not have been wiser to have continued with the slaughter policy and met the cost from some form of insurance levy on the industry.

Other risks can be as great a loss factor to the individual broiler producer.

A rat gaining entry to the broiler house, or an owl, can kill as many birds as disease. An owl getting into one of these houses was so frightened that it flew backwards and forwards with the birds piling up on each other in equal terror, until 800 birds were lost through suffocation. Any fright of this kind can be especially dangerous when, at the end of their short lives, the birds have $\frac{1}{2}$ to $\frac{3}{4}$ sq.ft. per bird and any panic away from the entrance means a piling up at the other end of the house.

Misfortune too can play havoc with profits. Apart from fowl pest there is always the risk of mechanical aids going wrong. For example, any breakdown of ventilation fans can kill all the birds at one go. A broiler manager entered his house one morning and noticed that all was curiously still and quiet. His birds were all dead, his ventilation fans having stopped working. The temperatures of the houses can be controlled upwards by thermostat but not downwards because refrigeration units would cost too much to instal. 'I just cross my fingers and hope for the best during a heat-wave,' the manager told me. 'More than a day or two of very hot weather can kill them.'

In the early days of broiler production several fortunes were made, but as in other walks of life, the more who enter the industry the smaller the profits. Quite often in recent years there has been a great glut of broilers on the market. Edward Trow reported in the *Daily Express* (12th February 1962):

> Chickens by the thousand are locked up in butchers' and wholesalers' stores today.
>
> Reason: the men in the trade can't sell them. Not for years has there been such a drop in poultry buying, although chickens are cheaper now than they have been for three or four years.
>
> … My comment: it could be that people want boilers for taste instead of broilers for no taste.

The supermarkets have taken to broilers with great gusto. They signify a new affluence status. They are given away when a new store opens, or sold as loss leaders to lure customers inside. Well and good. But the customer gets used to these low prices and it is difficult to wean her back to realistic prices. 'Tell your producers not to follow our pattern unless they want the same cemetery that we are entering,' warned Mr Norman Beeston as he left the States after a fact-finding tour, in 1961:

> 'Broilers', he said, 'are reaching the consumer at 10 cents a lb., four cents below their cost of production. They told me it was the culling season but if all that I saw and was told is correct, their culling season extends over a remarkably long period.'
>
> Pointing direct at the dangers which would beset British producers if the American pattern was adhered to, Mr Beeston said: 'The American broiler industry is producing an excess with which they do not know how to deal. Production is not their problem, marketing is.
>
> 'Employment of the broiler as a loss leader is, to say the least, a suicidal move,' he continued. 'This sort of thing has only to be carried on for long enough to convince the housewife that the cut price is the correct price. The result is that the retailer is selling at a price which could last for all time.' (*Poultry World*, 29th June 1961)

Profits have been as low as $\frac{1}{2}$d. a bird. Many small producers, relying on credit from their banks, have been forced into bankruptcy when the credit squeeze came and broiler production outstripped consumption. Obviously the higher production soared the more difficult it was to sell and the prices fell and fell. To counterbalance this some more enterprising producers have evolved their own methods of making a profit. One farmer has taken a stall at his local market and finds he can get rid of a large number of 'fresh' birds every week, others sell at their gates, or from door to door in nearby districts, one has even started a flourishing business of a mobile barbecue, delivering the chickens to housewives still hot and 'basted in butter and sherry'. No mention is made to housewives that the chickens are broilers for the name 'broiler' is unpopular. In 1961 the British Broiler Growers' Association found it wiser to change its name to the British Chicken Association.

This then is the broiler industry, vast and struggling, a business rather than an agricultural enterprise as we think of agriculture. And the chickens, after nine to ten weeks in these dim, enclosed houses, reach their required weight of $3\frac{1}{2}$lb. and are caught, crated and sent to the 'packing station'. The fact that after all this effort the chickens so produced are absolutely tasteless, is a matter which will be dealt with in another chapter.

# Poultry Packing Stations

We made our way down to the innocuous looking factory-like building. On the outside of the vast shed with its great sliding doors hundreds of crates were stacked high against the wall, just as they had been unloaded from the lorries.

White overalls and wellington boots were sent for to protect us from the blood in the slaughter room and then we went into the shed. More crates were stacked up the inside wall, twelve birds to the crate, with a grandstand view of all that went on: the noise and bustle, the music-while-you-work, the harsh light, the throb and clatter of machinery. They had come, these ten-week-old birds, from their dimmed out, hushed and cossetted environment, protected from any unusual noise or disturbance. They had been bundled into crates and onto lorries and for the first time had seen God's own sun above their heads. For the first time and the last time. They had come to the end of their journey.

Not that they could have been in an ideal frame of mind to savour the full benison of their brief period in the light. As in all slaughtering, the birds are advisedly starved for twelve to sixteen hours before they reach the packing station and they are apt to spend the best part of a day in their crates after they reach it, before their turn comes. During this time they get neither food nor drink, because any undigested food is waste and can impair the keeping quality of the carcase in the deep freeze. In this business a half ounce of food per bird is big money, it could mean the difference between a profit and a loss.

Their time is nearly run out. Taken out of their crates they are suspended by their legs on a moving belt, gently because they mustn't be

 © J. Harrison and J. Wilson 2013. *Animal Machines* (Ruth Harrison)

frightened or they would not defeather so well! The time taken to reach the slaughterman varies between one and five minutes according to the layout and speed of the conveyor belt. As they move along their beaks open and shut, mutely, in what has all the appearance of fear, but I was told that chickens are dim creatures and have not the slightest idea of what is happening to them. When they reared up and flapped their wings a gentle pulling down on their necks assured quiet hanging and on they went.

I noticed that there was a time when for about half a minute they passed within a few feet of the conveyor belt moving in the opposite direction bearing the other chickens, still shackled, but already defeathered, and that it was at this time that they showed most symptoms of fear. Was I deceiving myself? Was it a coincidence? On a later occasion I asked the naturalist Konrad Lorenz whether he thought they would realise what was happening and he replied: 'According to my experience, chickens do not "understand" the situation when their fellows are being slaughtered or lying dead, it certainly would not increase their suffering. Conversely, I am convinced that cattle suffer the torments of extreme terror when entering the slaughterhouse because they do smell the blood of their own species.' A distinguished poultryman, on the other hand, seems to differ. Murray Hale writing in *Poultry World* (5th October 1961) said:

> … The awareness of danger is the only thing that enables a species to survive in the wild, and no amount of domestication can eradicate these deep-seated instincts. Genetically they are not only dominant, they are paramount.

and Mr Fry of Reading wrote to *Poultry World* (22nd June 1961):

> … All the consumer wants is a clean, well-fleshed wholesome looking bird. Where this is not achieved I think the packing station may be more to blame than the grower.
>
> In a good many instances birds are piled in crates on the floor in full view of their colleagues being slaughtered and there is no doubt but that they know what is in store for them. This frightens them and a scared tensed-up bird does not pluck well, in my opinion, because the grip on its feathers is increased.

Some packing stations have stunners to use before the birds have their throats cut, some do not. Some have stunners but do not use them. The one I visited was one of these latter. 'They do not bleed properly,' said the manager, 'it is much quicker this way, and kinder too.' I watched the birds having their throats cut and disappear, flapping wildly, into the bleeding tunnel, to reappear a minute later still flapping wildly, to go into the scalding tank. 'They are dead before they go into that,' said the manager reassuringly. When the birds came through the scalding tank they were limp and dead. They then went through a defeathering machine and at this point were conveyed past the live birds and through a gap in the wall into the evisceration room. In the evisceration room the conveyor belt moved slowly over a bench running the length of the hall, with a gulley underneath for waste. Along this bench stood dozens of white coated young girls and men and as the conveyor belt moved they each did their 'piece work'. One weighed the bird, the next cut off its feet and re-suspended it on the belt this time by its head, the next slit it open and so on. When it was thoroughly clean, inside and out, it went through a chilling machine and the conveyor belt then passed into a third room where final shaping and packaging into polythene bags went on. Here any bruised or battered birds were removed and cut into 'chicken pieces', the disfigured parts being removed first. The polythene packs are piled into wire trolleys which are duly wheeled into the deep freeze rooms for storage until a sale is made. At this time the frozen packs are put into gay cardboard boxes showing some happy picture to attract the housewife's attention in the supermarket or store.

There was a general air of contentment amongst the workers: the girls hummed and the men chatted. Even in the slaughter room people walked casually around, pushing the birds to one side, like curtains, if they were in the way, and then if the birds seemed disturbed, pulling their necks to quieten them again. The whole process, from crate to deep freeze takes between eighteen minutes and upwards of an hour depending on the rate of the line.

'Do you eat broiler chickens?' we asked the manager. Back came the prompt reply, 'Good lord, no!' I later had cause to visit an abattoir and asked this same question of the superintendent: 'Can you bear to eat meat after watching the animals being killed?' 'I can't eat young animals,' he said, 'but watching the bigger animals being slaughtered doesn't affect me.'

Onto the conveyor belts go ducks, unused to being handled upside down, turkeys and rabbits. This packing station handled, at that time, 1,500 birds an hour. It would be rated as a small unit these days. One wondered, watching the slaughterman's knife in action, the blood spurting onto him, what happened when he sneezed, or had to wipe his nose, or had a tickle, or if anything diverted his attention from the birds for a moment. Did they pass by and go into the scalding tank fully alive? This close timing of a bird every two seconds cannot allow for any human weakness. One wondered also about the slaughterman. Fifteen hundred birds an hour is a lot to kill skilfully and efficiently for eight hours a day; was it possible that after a time he would be concerned only that bleeding took place and not with the suffering of the birds? 'We have no difficulty in getting slaughtermen,' I was told.

Mr Peppercorn, Chairman of the British Chicken Association (*Farmer and Stockbreeder Supplement*, 30th January 1962) predicts:

> The packing stations of the future are likely to be considerably larger than most packing stations are today because they can do the job more cheaply. I don't think anyone yet knows what will be the most economic size

for a packing station in this country, but it would be fairly safe to say that it will be a minimum of 30,000 a day, and may well be substantially larger than this.

Already the larger packing stations are handling 3,000 to 4,500 birds an hour.

With more than 200 million birds passing each year through the packing stations, there is still no legislation for protecting them from having their throats cut while fully conscious. This was due in the first instance to the difficulty of finding a way to kill the birds humanely. It was at first hoped that gas would be a good method, but the flapping wings of the birds as they went through the gas chamber caused a loss of $\frac{1}{2}$lb. gas per bird, too expensive a method for the industry. The Humane Slaughter Association, in association with veterinary surgeons, decided that electrical stunning was the most humane alternative.

A high voltage stunner was at first produced but the factory inspector came in and tried to stop it because it was dangerous to use. It also killed half the birds and only paralysed the rest. They had to stay alive long enough for their hearts to pump out their blood. Then Mr Cotton of Cope and Cope Electronics, who had invented this stunner, started his research all over again to discover how he could produce a stunner with voltage low enough to stun these young birds without killing them. It had also to be applied long enough to keep them unconscious for 90 seconds, the time taken to reach the slaughterman and be bled, without regaining consciousness. Timing and skill became of prime importance. He finally invented a low voltage stunner which seemed to fulfil all the necessary requirements and, with Miss Sidley of the Humane Slaughter Association, and Mr R. A. Wright, M.R.C.V.S., of the Houghton Poultry Research Station (which is run jointly by the Ministry of Agriculture and the Animal Health Trust), he ran a series of tests to prove its efficiency both from the point of view of humaneness and from that of good bleeding and carcase value. The Annual Report of the Humane Slaughter Association 1959–60 itemised Mr Wright's conclusions as to the success of the stunner:

(*a*) safety in operation
(*b*) apparent complete insensibility in birds
(*c*) complete relaxation of voluntary muscles (except spasmodic wing flapping) which renders jugular severance easy
(*d*) effective bleeding
(*e*) mechanical plucking is easier
(*f*) fractures of limbs rarely occur

The Report also gave Mr Wright's comment on the killing of birds without prior stunning: 'I have no hesitation in considering jugular severance without prior stunning as being grossly inhumane as birds were obviously fully conscious and in great pain for some appreciable time.'

Having established the low voltage stunner to be effective both in kill-ing the bird humanely, and in allowing good bleeding, the Ministry were approached to make its use compulsory and they sent one of their veteri-nary officers to a modern plant to check up and give an opinion on it.

> Miss Sidley was asked to attend a meeting of ministry officials when the veterinary officer concerned said that in his opinion one in every nine birds appeared to show a certain amount of eye reflex and that he was not prepared to recommend the method to the Minister until further research work had been carried out to prove that eye reflex is not associated with consciousness.
>
> The research work is being carried out at Houghton Poultry Research Station and a report will be submitted to the Ministry as quickly as possible.
>
> A number of eminent physiologists have stated publicly that corneal reflexes are absolutely no criterion of the existence of consciousness.

I was informed by the Humane Slaughter Association, from whose 1961 Report the above quotation was taken, that of the birds which have their throats cut in full consciousness two out of five go into the scalding tank alive.

When one considers that well over 150,000,000 broiler chickens and probably half as many again battery chickens and intensively housed layers go through the packing stations every year, one can see how many millions suffer this fate, even if, as we are led to believe, only a handful of packing stations operate without stunners.

Mr Cotton wrote to me in June 1961:

> I find on checking our figures that we have supplied quite a large number of high voltage stunners; at the time of writing not quite so many of the low voltage type, but I feel that time will show that the low voltage will prove the answer. High voltages are dangerous to use and more likely to kill or paralyse the bird.
>
> Generally speaking, as far as low voltage stunning is concerned, if the Station is a small or modest one (up to say 400 birds an hour) then if they have bought a low voltage stunner they are probably using it, for this is about the speed at which low voltage stunning can be effected. Where the plants handle numbers greatly in excess of these, then while they have bought a stunner, they are probably not using it as the lines move too fast.

When this low voltage stunner was finally perfected, it might have seemed a happy solution to meeting the humanitarian requirements of the poultry packing stations, but in fact it proved virtually useless because the larger packing stations had meanwhile increased the speed of their con-veyor belts to take upwards of four thousand birds an hour and this speed precludes adequate stunning of the birds with these methods. One can-not help wondering how many packing stations considered meeting the new problem by the employment of more than one operative for stunning, rather than adopting the high voltage which can merely paralyse the bird and so nullify the humanitarian aspect of stunning, or allow the birds to go through to the slaughterman fully conscious?

Obviously packing stations are out to make as much profit as possible on the chickens. Is it that they resent having to pay a highly skilled man, or men, to pre-stun the birds? There are, of course, other stunners on the market. There is the electric knife, for example, which stuns as it cuts. And there is the new Maywick stunning box, through which the line moves, brushing the chickens over a sheet of metal charged with electric current, where they are stunned, while the slaughterman waits at the other end of the compartment and bleeds them as they come out. This apparatus is high in initial cost, but it eliminates the employment of an extra man, and would seem to be foolproof. There is also a similar Danish one which has been brought over here and can handle 4,000 birds an hour.

Why, one asks, can't the birds have their necks dislocated in the old way, a method which when well done was painless, instantaneous, and which, it seems, caused all the blood to drain into the neck and presumably left the carcase the desired white colour? Well, it appears that, tough as it is on the chicken, there is some doubt about *all* the blood draining away when the heart is not beating to pump it out, and the carcase would not have the same keeping qualities in the deep freeze. The carcases sometimes have to be stored for many months during periods when the public is less fond of its 'spring chickens'. There are other factors which affect carcase quality.

Murray Hale wrote in *Poultry World* (5th October 1961):

> … Broilers, crammed into crates, bumped and chased and bruised into a state when they lose moisture, are generally a grade poorer than when they were picked up. If you get down-grading slips from your packer, take a look at the way the birds are handled from the time they are caught on your farm, in the lorries, and at the killing line.
>
> There is far too much needless rush, too much labour saving that is so costly in the invisible depreciation of reduced production. I would not worry if a man could not read or write so long as he understood how to treat his stock at all times.

Poultry packing is backed by constant research just like other branches of the industry. Machines for defeathering, chilling, freezing, scalding, and evisceration are being constantly made more automatic and more thorough. Thought has gone also into a new tranquillising drug to put into the water trough before the birds are caught and so ensure gentler handling and avoid the hysteria and feather flying normally involved in catching 10,000 birds for crating – no mean task!

Robin Clapham (*Farmer and Stockbreeder*, 29th August 1961) tells us of methods other research scientists have undertaken to utilise some of the 'waste' from packing stations:

As one instance I select the current investigation into the feeding of blood, feather and offal meal from broiler packing stations to chickens. How cannibalistic can you get? Yet the amazing thing is that the birds did jolly well on it, apart, that is, from the report's observation that the 'disadvantage at low levels of 5 and $7\frac{1}{2}$ per cent in these diets was depressed feed intake...'.

... But there is also serious work afoot on the nutritional value of hydrolised poultry manure for broilers. Based on chemical analysis, it was shown that about one-half of the crude protein of hen manure and one-third of that of broiler manure used in this protein study existed in fact as true protein....

And *Farmer and Stockbreeder* reports (10th October 1961):

Is it ridiculous to be thinking of birds without combs and wattles, and even without wings? asked Mr Derek Kelly, general manager of Associated Broiler Breeders Ltd., at the Chick Producers' Association conference in Folkestone, last week. ... Packing station profits were susceptible to the conformation of the bird, and the ratio of edible meat to viscera. Whilst total live weight was important, every gramme of meat gained at the expense of a gramme of offal could be considered a free bonus....

'Imagination could run riot when considering ways and means of reducing the wastage in processing from live weight to oven-ready,' said Mr Kelly. 'The meat/viscera ratio, length of shanks, and neck length are all now receiving attention.'

Dr Abbott, of California University, had been experimenting with chickens 'in a novel environment'. These birds were featherless. Although admitting that it might sound fantastic to suggest this today, Mr Kelly thought that in 10 years time it might be considered the logical approach to the elimination of one stage in the processing plant.

In May 1961 *The Veterinary Record* commented on the lack of regulations for inspection in poultry packing stations and the possibility of disease spreading amongst flocks and to man himself:

In this country poultry intended for human consumption is subject to the usual regulations governing all meat, etc., deposited for sale. But the regulations and licensing controls relating to the slaughter of other animals do not apply to poultry. Whereas a substantial percentage of red meat undergoes inspection prior to release for human consumption, most poultry escape any form of regular inspection, except under local by-laws and regulations of certain of the larger cities. Routine inspection of poultry carcases, instituted in 1936 at Smithfield Market, revealed over 60 per cent to be diseased in one way or another. That figure may be misleading, as at that time a much greater proportion of the carcases inspected would have been those of old fowls as compared with the present day when the emphasis is on the younger bird; nevertheless, there is no doubt that an appreciable number of diseased carcases could reach the public if standards were allowed to fall at the different processing factories....

Consumers of poultry and poultry products are entitled to a clean and wholesome article processed under strict sanitary conditions, but what are the real risks to the consumer if no attempt is made to produce and market disease-free poultry?

The normal procedure where meat inspection of poultry is carried out is to reject a carcase as unfit when there is any condition present which might affect its appearance and palatability. It is often difficult, if not impossible, to diagnose the actual disease condition present; and whether or not it is transmissible or harmful to man is, therefore, only a secondary consideration. There can be no question that this approach is correct. The carcase of a sick bird rarely cooks satisfactorily and the flesh is liable to be tough and flavourless, if not actually tainted.

Furthermore, strict standards are advisable when handling live poultry and carcases for reasons other than those associated with actual human consumption of the product. There is the possible risk to those handling the birds and carcases during processing, and also a danger of spreading disease through the poultry industry itself as a result of the use of dirty crates, and distribution of offal from infected carcases.

Conditions do not seem to have improved much since, for in 1962 the Plant Committee Report on Fowl Pest commented:

We were shocked by the very low standard of hygiene at some of the premises which we visited where poultry are slaughtered and we have been surprised to learn that comparable conditions exist elsewhere. Local authorities have certain responsibilities for hygiene. We recognise that standards at many packing stations in this country are high, but often the nature and structure of the premises are severely limiting factors. We think that the centralisation of poultry slaughtering ought to be encouraged in order that premises may be used where a good standard of hygiene can be readily achieved and maintained. It was of considerable interest to us that great care was taken by the responsible authorities in the Netherlands, U.S.A. and Canada to secure hygiene in poultry slaughterhouses and we look forward to the day when the efforts of the Ministry of Health and local authorities ensure the maintenance of comparable conditions here.

We are informed that a Code of Practice on Poultry Dressing and Packing has recently been published but at present there is no official poultry meat inspection scheme such as applies at most poultry packing stations in the U.S.A. The persistence of Newcastle disease in this country may be attributable, in part, to the spread of infection from the carcases of birds slaughtered in the early incubative stages of the disease. It is not likely that such birds could often be recognised at the slaughter-house by meat inspectors and we do not wish to press for this service as a measure for the control of Newcastle disease though it could provide an indication of the incidence of respiratory infections in general.

*Registration of Slaughterhouses*

Manchester Corporation obtained power in the Manchester Corporation Act, 1954, to register poultry slaughterhouses in the city. These powers are being used and we understand that the Corporation consider that they are valuable. Local authorities must know the whereabouts of poultry slaughterhouses if those of them that have responsibilities under the Diseases of Animals Act are to be in a good position to enforce the regulations. *We recommend that the desirability of making arrangements in all local authority areas similar to those in Manchester should be examined by the authorities concerned.*

At the time of going to press the Ministry have told me that there is still no legislation requiring registration of poultry packing stations in order to enforce inspection of carcases as being strictly safe for human consumption.

# Battery Birds

Most farms have poultry, and most of their flocks of laying hens are small, consisting of five hundred birds or less, but as specialisation has taken hold of the field, flocks have tended to become larger, and now, in England and Wales, it is estimated that a third of all eggs are produced in flocks of over five hundred birds, and a sixth in very large flocks indeed. Several egg tycoons have flocks of a hundred thousand birds!

Chickens, like other animals, are fast disappearing from the farm scenery. Only 20 per cent are now on range, whilst 80 per cent have gone indoors. What has caused this trend?

Probably the most important factor is that birds lay fewer eggs when they are cold, their energy goes into keeping warm. Because the bird, left to itself, will dissipate its energy keeping warm and foraging, the farmer who wishes to conserve this waste and channel all its energy into the single function of egg laying, deprives it of this natural pleasure.

Free range hens always have access to shelter, be it in an outhouse or barn, or in a range shelter purpose-made for them. So that if they are seen, as one 'intensive' farmer scornfully put it, 'scratching around in the mud, open to the wind and rain', could it perhaps be that they like to be outside? In the *Oxford Mail Farming Supplement*, 16th February 1962, a poultry farmer with a thirty-year run of major awards in laying trials behind him, confirms this:

> ... (The hens) have the choice of snug indoor conditions on deep litter or the fresh air of grass range.
>
> But they have a preference for the rugged outdoor life even in winter. Not even a fall of snow daunts their determination to range over the ground.

© J. Harrison and J. Wilson 2013. *Animal Machines* (Ruth Harrison)

During last month's snowfalls the fertility rate was well above 90 per cent – which destroys the theory that it is no longer economic to keep breeding poultry outdoors during winter.

Until recently nearly all birds were reared to point of lay outdoors and, even now, a very large percentage still are. This applies also to breeding hens. It is considered that outdoor rearing makes for a much healthier, hardier bird, tight feathered and less prone to disease, a safeguard against the strain of its life in the intensive unit into which it goes at point of lay. *Poultry World*, 8th November 1962, reported on a farm where an experiment was tried with two types of building for breeding hens, the first a window less, controlled environment house, and the second little more than a large shelter open at both ends to wind and weather. It was found that at point of lay the birds from the first house 'looked anaemic and their body size was significantly smaller than similar birds reared under the other systems. Performance in the laying houses afterwards also seemed to be adversely affected, the birds peaking at a 10 per cent lower production rate than the others. The only encouraging aspect was that their egg size was superior to the other pullets, and adult mortality was no worse.' Of the results in the second house, it was stated: 'In this fresh air environment pullets rear exceptionally well, feathering tightly, growing rapidly even in cold weather and experiencing no health problems at all during the whole period.'

Many farmers have faith that this is in the end the best way to rear chickens. Here is a report from another farm:

Laying pullets destined to spend their productive lives in a super-intensive environment of wire floors or slats, need stamina and bubbling good health if they are to express to the full their inherent 250 eggs-per-year potential. This, to Mr J. A. Reid's mind, means that fresh air, clean ground, sunlight and grass during rearing are just as important as they ever were – perhaps even more so these days....

Labour-wise, Mr Reid could slash his rearing costs considerably if he sold off his folds and invested in a large intensive rearing house. But he will not hear of this. His firm belief is that the more intensive egg production methods become the more will there be a need for the layers to build up fitness and stamina in the fresh air during the vital growing period. (*Poultry World*, 24th August 1961)

'Can I rear pullets to point of lay in cages?' asked an inquirer to *Farmer and Stockbreeder*, 11th April 1961.

'I would not advise it,' was the reply; 'free range is not necessary, but I consider that access to fresh air and sunshine is....'

'Those people who are breeding generation after generation of birds in battery cages,' commented a farmer in Oxfordshire (*Oxford Mail Supplement*, 16th February 1962), 'will have to do what they are doing in America – that is, come back to the people who are using the outdoor system. I have been told they are getting trouble after so many generations of intensive breeding.'

But if a bird does not lay as many eggs in winter as in summer it is not 'paying for' its feed, is making no profit for the farmer and is consequently not an economic proposition. So, like it or not, indoors it goes on the 'business farm'.

There are many intermediate stages towards intensivism between free range and battery cages, but the four chief forms of intensive housing are (a) deep litter, (b) wire flooring, (c) slatted flooring, and (d) battery cages. Of these the only one which confines the bird completely is the battery cage, but let us see what these other terms mean.

Most of the more recently built intensive houses are of the same basic design; long, wide, windowless sheds, with air vents down the sides and a row of extract fans along the apex of the roofs, much the same as the broiler houses. Inside, the floor of the building will be of concrete, covered with a nine-inch layer of litter, or there may be a raised wire mesh floor or raised slats through which the droppings can fall onto the slab below. There are variations on the theme. Floors may be partially wired and partially slatted, or partially wired and partially littered, and so on. All the sheds will have troughs or hoppers for feeding and a piped water supply. All will have nesting-boxes down one side with provision for gathering eggs from outside without disturbing the birds. There will be broody coops and places to perch.

Within this context the birds are free to roam at will.

Recommendations for stocking vary between one and four square feet per bird, but emphasis is always laid on the fact that overstocking can cause vices. As in the farmyard, there must be room for birds lower down the social scale to live in peace with birds higher up the pecking order. This rigidly established order is maintained through all the activities of the flock and, where facilities are inadequate, as for example in a lack of trough space, birds of a higher pecking order will guard the trough against the lower birds. Apart from a danger of malnutrition among the weaker birds, the more extreme conditions of overcrowding conduce towards the vices of feather-pecking and cannibalism and it must follow that the weaklings have a pretty miserable and debilitated kind of existence. The struggle to establish the pecking order begins at about the age of ten weeks and, in an attempt to forestall the vice of pecking, it is common practice to have the birds de-beaked before they reach this age.

There are, of course, many old converted farm buildings used for laying stock, and some have the advantage of big windows letting in the sun and fresh air, whilst having the presumed disadvantage of no control of temperature or lighting.

Of the forms of intensive rearing of laying poultry the battery-cage system confines the birds most closely and I shall be devoting the bulk of this chapter to a consideration of this system.

Keeping birds in cages is no new idea, and even keeping hens in battery cages dates back fifty years. But it is only in the last decade that battery cages have achieved the popularity associated with them today.

In 1911 Professor J. Halpin of Wisconsin University, U.S.A., kept hens in cages of thin wood, with wire at the front and top and a door in the front panel. These cages were arranged in three tiers. But it was not until 1924 that laying cages recognised as such were adopted in America, and only in 1930–31 that they were manufactured commercially.

In this country battery cages were pioneered by a Mr Winward, a farmer in Lancashire, in about 1925. The framework was in timber, the floors and divisions in wire. They had individual wooden food troughs and a jam jar to each two cages. It took this farmer two years to build two thousand cages, but these were then in continuous use until 1940. In 1930 cages were manufactured commercially in this country also, and from then on the industry grew steadily. Development was brought to a halt during the war years and not until the end of foodstuffs rationing ten years ago did it really begin to snowball to its present enormous proportions. Dr Blount of British Oil and Cake Mills estimated from sales figures of battery cages that in 1951 there were three million cages in use. Today, half the intensively housed laying birds are estimated to be in batteries, that is 40 per cent of all layers. In 1961, a 'poor' year it seems, there were seventy million laying birds in the United Kingdom, so there were some twenty-eight million birds in battery cages.

What is it, apart from the fact that the birds lay a few more eggs in the winter, that attracts some farmers to battery cages, as opposed to other intensive methods less restrictive to the birds?

The chief advantage to the farmer is again economic. In theory he can cull, or kill and replace those birds which do not pay for their food by laying enough eggs. I say 'in theory' because when birds were only one to a cage this must have been easy to spot, with charts above the cage on which eggs could be entered up, but now that the squeeze is on and three, four or even more birds are crammed into one cage, it must be much harder work to spot the 'passengers'. I am told that an experienced poultryman can tell by inspecting a hen whether or not she is laying. Be that as it may it becomes statistically more complex to pick out sub-standard layers as the number of birds per cage rises, and this fact must vitiate the advantage the system had in its original form of one carefully charted bird per cage.

Another attraction, as in all intensive rearing, is that he can stock much more densely than by using other methods. He can set up on a plot of land just able to take his battery house and run it as a unit in itself, completely independent of its surroundings. He is also theoretically independent of weather conditions. This theory, too, can be demolished by a winter of the severity of that experienced in 1962-3. The *Evening Standard* reported during that winter's blizzards:

> Near Honiton, Devon, local farmers were trying to cut a way through 16 ft. drifts to poultry-keeper Edric Berry, cut off at Dunkeswell.
> He has 30,000 laying-hens.

Said Mr Berry: 'They had their last feed this morning. I need three and a half tons of food a day. The situation is desperate. We've been cut off since Friday – and the hens are still laying like mad. I've got nearly a hundred thousand eggs lying around everywhere. It's a nightmare.

In the battery unit automation is exploited to the full. The photographs in Figure 20 show that modern banked cages are made in large units which have the appearance of immense machines, which indeed is what they are. The cages are ranked one above the other, three, four or even five tiers high. Units are placed in rows end to end, back to back, until they fill the space available, the passageways between tiers being just sufficiently wide for servicing.

Food troughs run the length of the cages, kept continuously supplied by a conveyor belt from a hopper at one end. The hopper needs refilling only once a day or less frequently. Water is laid on in a second continuous trough, or by means of a piped feed with individual take-off valves at each cage. The technical arrangements compel admiration.

The bird stands on a wire grid which has a 1 in 5 slope from back to front so that the egg rolls away into yet another rack in front of the cage, where it will not be soiled by droppings and is nicely calculated to be just too far away to be eaten by the bird.

The engineers can go further than this. Systems have been devised which transport the egg to central collecting points, thus obviating the one remaining regular chore in the battery house. However, the objection to this refinement is that no check on the non-layers is then possible because the tell-tale eggs have disappeared from the scene. Whilst it is mechanically feasible to devise means of counting the eggs as they fall from individual cages, it would be extremely difficult to eliminate the soft or shell-less eggs which are sometimes produced and, all in all, this particular refinement has not yet been much used.

The droppings fall onto a tray which is provided beneath each row of cages and which is automatically 'squeegeed' off at intervals into a pit or container. A moving belt may be used instead of a fixed tray, and in some of the larger installations lateral belts carry the manure to a central collecting point, from which it can be disposed of by arrangements individual to each farm. Some farms run the effluent direct into the local sewer, some direct onto their farms when they have a general farm, some sell it to market gardeners. But in all cases there is not much labour attached to its disposal, except perhaps in urban areas.

In the windowless houses there are automatic switches which can be set to give the birds the exact amount of light they need for maximum egg production and a complicated system of lighting patterns exists to guide the farmer to this happy state.

For a long time only one bird was housed in each cage, then two were tried together. Mortality was no greater, and the birds even seemed to enjoy the companionship. Profits were a little higher because there was a saving on capital expenditure. Then three birds were tried to a cage, some even tried four to the ordinary 15 or 16-inch cage. Results in terms of profits seemed fairly successful:

A correspondent in *Poultry World* wrote (11th October 1962):

> At first, two pullets were put in each $13\frac{1}{2}$ inch cage, but they appeared to have so much room to spare that a trial was carried out to compare performances with three birds to a cage.
>
> The differences in egg production, feed conversion and mortality were so insignificant that all the cages now hold three birds.
>
> Such a density of stocking – only $\frac{2}{3}$ sq. ft. per bird – demands a highly efficient ventilating system in the house....
>
> Greater margins of profit from laying stock could result from housing three birds to a cage in future ... (stated an article in *Farmer and Stockbreeder*, 1st May 1962).
>
> 'Return on capital invested is always likely to be greater with three birds per 15-inch cage than with lower densities of stocking.'
>
> It is worth noting that extra mortality experienced with three birds in a cage was about two birds per 100 pullets housed. It occurred in the first two or three months of lay and was chiefly attributable to vent pecking and other forms of cannibalism.

and *The Farmer's Weekly* reported of a farm where:

> Using $17\frac{1}{2}$ in. cages in four tiers and three blocks, ... (he) houses 1,728 birds with an allowance a bird of $\frac{5}{8}$ sq. ft. This cuts the capital cost of the fully furnished housing to £1 a bird....
>
> A further battery house of similar size and construction is now being erected which will house 2,304 birds (four to a cage). Each bird will get 0·45 sq. ft. of cage space. (13th October 1961)

One firm are making production comparisons with 'one bird to a $9\frac{1}{2}$-in. cage, two to a 12-in. cage and three to a 16-in. cage. Other possible combinations are four to a 16-in. cage or four or five in a 24-in. cage.' And *The Smallholder*, 22nd July 1961, advised:

> Finally, battery owners have to decide how many birds to keep in a cage, and how much space is required per bird. The more layers that can be kept in a battery house, the greater the economy – but this saving can easily be nullified if, as a result, production per bird drops.
>
> Experience both in this country and in America shows that production per bird does not vary no matter how many birds are put together; mortality, however, increases considerably when more than two are caged together. Twin-bird cages therefore seem to offer the most economic production.
>
> Some producers have found that two small hybrids will lay well in a 9-in. cage, but it is more usual to allow them 11 or 12 in., and 13 or 14 in. for two heavy pullets.
>
> In Britain (reported *The Farmer's Weekly*, 9th March 1962), where the laying cage was pioneered, operators seldom put more than 1, 2 or 3 pullets to a cage.

But in America – where the laying cage is gradually creeping in – there is none of this fiddle-faddle. The minimum-sized cage popularly used is of 20 in. × 30 in., holding 10 layers, but the more common size is 3 ft. × 4 ft. or 3 ft. × 5 ft. with 15 to 25 layers in it.

But a Special Representative, *Poultry World*, 6th September 1962, warns:

> Overlooked by many poultrymen, however, is social behaviour. The peak order could be almost as serious in multi-bird cages as on wired or deep littered floors ... it can hold a flock back from peak production by at least 10 per cent....

It appears that the two ways of dealing with feather-pecking and cannibalism in multi-bird cages is either by de-beaking or by reducing light intensity. He quotes Dr Jones of Thornbers:

> 'We find a much lower intensity under controlled environment is equally efficient. We use between three-quarters and one lumen per square foot.' Lighting is red but it is not known yet whether it is the colour or intensity of light which does most to control the vice of pecking....

One man and a boy can look after as many as 15,000 birds, when they are fully automated in a battery unit, but they cannot give the personal attention and interest to be had in small units. Eric Baird thinks that this should not be regarded as of prime importance, but rather as a 'plus factor' (*Farmer and Stockbreeder*, 14th March 1961):

> It has been said lately that small units score as a result of personal attention to detail. This I don't deny. But at the same time it is important to appreciate that where hens are just one of several or many lines of production too much faith should not rest in this sometimes doubtful benefit.
>
> It is better these days to ape the large-scale units and make your poultry outfit foolproof so that the stockmanship factor is a plus one. That way your profit margin should be reasonably assured. It follows that the process of doubling-up the profitable unit is a good policy;

whereas a writer in *Poultry World*, 17th May 1962, feels that mechanisation cannot in any way take the place of stockmanship:

> Generally speaking the mechanised labour-saving device will prove most profitable on those holdings where the employer of labour knows that the staff engaged on the management chores have little or no interest in the stock. And this does happen all too frequently nowadays.
>
> Stockmen in the hen house are (more is the pity) something of a rarity, and – where they are still to be found, mechanical aids, it has to be admitted – are likely to be conspicuous by their absence.
>
> That is not to denigrate out of hand the mechanical innovations that are likely to become more and more available. Rather is the point being made that it will be a major error of judgment to suppose they can replace the basic know-how of the poultry job.

The automatic feed and water supply is a case in point. It is not being used to advantage if the operator fails to notice that for one reason or another the birds for which they are installed have fallen below their normal intake.

Your stockman will take notice of this fact immediately, and by so doing is possibly getting advance notice that something is not well with the stock in his charge. Outbreaks of disease can be checked by his vigilance before they gain a firm foothold.

It cannot, therefore, be put forward as a justifiable claim that mechanisation is, factory-wise, the large-scale producer's positive insurance that eggs, or table birds, will be produced more cheaply than on the smaller plant, where, possibly, family labour is, of sheer necessity, of a more conscientious and hard-working character.

Anthony Phelps, writing in *Poultry World*, 8th February 1962, wonders why producers should feel that intensity of stocking is the best way to reduce their costs:

It is a little difficult to understand why housing invariably receives first attention when such economies have to be made. It represents only a relatively small proportion of the total cost of producing eggs, as these typical (and actual) laying flock costings show:

| Production Cost Items Per Bird | Cost | | Percentage of Total |
|---|---|---|---|
| Rearing to p.o.l. | 13s. | 11d. | 28 |
| Feeding from p.o.l. | 27s. | 0d. | 54 |
| Depreciation: | | | |
|    Housing, equipment | 4s. | 2d. | 8 |
| Labour | 2s. | 10d. | 6 |
| Electricity, water | 1s. | 1d. | 2 |
| Overheads | 1s. | 0d. | 2 |
| | 50s. | 0d. | 100% |

From these figures it is clear that feeding and rearing are the factors which offer the most scope for economising and should be attended to first. On the average poultry farm the elimination of feed wastage and 'going bulk' will effect far greater savings than any housing economy, short of keeping the layers in open fields.

The disproportionate attention given to housing costs probably arises because poultrymen think in terms of saving on capital outlay rather than in terms of housing cost per bird per year.

To put 4,000 hens in a house designed to accommodate only 3,000 may save several hundred pounds of capital expenditure in the first place, but during the life of the building this only represents a saving of a shilling per bird per year.

There is no suggestion that this saving is not worth while, but, unfortunately, where population densities are concerned there are complexities. There is a reliable weight of evidence building up that increased stocking rates in laying houses depress the performance of the birds. In other words, reductions in housing costs can be false economies….

There has been no serious research in this country on the subject of laying performances at varying stocking rates, but several centres in America have looked into the question very thoroughly.

At most of these, the performance of layers with three sq. ft. of floor space each has been compared with birds at half the amount and in every case the crowded birds laid fewer eggs, had a higher rate of mortality, weighed less at disposal and needed more feed per dozen eggs produced.

For instance, at the University of Missouri, the pullets with three sq ft. laid at an overall rate of 67·2 per cent compared with 61·5 per cent from the crowded layers. In a 500-day hatch-to-disposal cycle this represents a difference of 20 eggs per bird, worth about 5s. 5d. at our average prices.

At the University of Nebraska, the more densely packed birds needed 0·4 lb. more feed per dozen eggs than hens at three sq. ft. The difference in output was similar to that at Missouri University.

On a basis of 18 dozen eggs, this means that the birds at three sq. ft. needed 7·2 lb. less feed than the others, a further cash advantage of two shillings.

Alternatively, the hens with plenty of space laid 241 eggs in a year on about the same feed intake needed by the crowded layers to produce 221 eggs.

Also at Nebraska it was found that mortality increased by a fraction under 5 per cent when population density was doubled. And at the Beacon Milling Company's research station it was found that the crowded hens weighed $\frac{1}{4}$ lb. less when scrapped.

Many other centres have contributed to this line of research and their findings confirm that these figures are typical….

There must be an economic optimum population density and much more research on this subject is needed – particularly in this country. Until then, producers would be wise to proceed with caution with even small reductions in the floor space allowance per bird.

This conception of the unimportance of capital cost relative to rearing and feeding costs applies equally to deep litter or battery houses and illustrates very clearly that concentration as such has only a marginal bearing on the economics of egg production.

Dr Rupert Coles, Ministry of Agriculture Chief Poultry Adviser, writing in *Farmer and Stockbreeder Supplement*, 29th January 1963, reviewed the situation:

So we come down to profit through high productivity as the major factor. Since it seems that increased output is so all important what can the individual do about this? More prolific stock is one obvious answer – but why the wide range in performance levels with the majority of birds today showing so little variation under standardised conditions?

Is it not that in the surge towards poultry 'factories' we have forgotten the bird is an individual living entity? Is it not because in the 'new world' we have brushed aside management as a 'has been' along with other traditions?

The Americans, whom we are so keen to follow, now seem to be rediscovering management. Floor space per bird is becoming greater; there is more individual attention. There are some who believe we are reaching the production limits possible through conventional breeding and that future progress will attend unorthodox methods aimed at controlling environment.

'Stimulighting' was introduced by the Americans a short while back. This was an increase in hours of lighting to stimulate more activity in the pullet and encourage her to lay more eggs; lights were put on early in the morning and continued late into the night so that at the peak of her laying period the bird could have more than twenty hours light a day. But a suspicion has been growing in the minds of poultry keepers over here that this over-stimulus has made for nervousness and 'vices' in the birds, and that maybe it is not such a good thing after all. So they have introduced 'one of the most important advances ever made in the management of intensively housed growing and laying stock', they have introduced 'twilighting', or the dimming of lights in the houses, so that where three lumens per square foot were previously recommended, we can now make do with hardly more than a glimmer of light, the birds spending their whole lives in perpetual twilight. But there is a drawback to this, and that is that whereas the birds go on laying eggs in the dimness, it is difficult for their attendants to see to do their jobs in the house. So many producers who have to spend some time in the houses have opted for red lights which can be slightly stronger without 'disturbing' the birds. And of course, we must not forget the economic factor – a large saving on electricity....

In 1961, however, despite the fact that layers were kept for only a year and it was considered by some poultry experts to matter less if pullets were reared all the way through in batteries without the advantage of free range and fresh air for health, a curious thing began to happen.

The sudden death of apparently healthy, strong pullets in battery cages is presenting research workers with a problem.

stated a *Farming Express* reporter, 14th December 1961.

The birds die of heart failure, but neither the cause nor a cure has been found....

Dr W. G. Siller, of the Poultry Research Centre, Edinburgh, thought that the birds were suffering from 'cage layer fatigue'. In the peracute form, he said (*Farmer and Stockbreeder*, 19th December 1961), 'the birds drop dead'. In the acute form 'there is prostration ... the birds will die if neglected, but if hand-fed or nursed they may recover after several weeks or even months,' and this, the article points out, 'is, of course, uneconomical on a farm scale'.

The birds which were most prone to this condition were White Leghorns. Post mortems showed the birds to be normal except for their bones being thin and soft, and eggs had been normal without soft shells. This led the geneticists to feel that it was not calcium deficiency which caused it. *Farming Express* went on to quote Dr Siller as saying:

'Only pullets confined to cages develop fatigue, while similarly bred and fed birds housed in pens are resistant. We think, therefore, that when certain strains of pullets are kept in cages for some time before starting egg production, the lack of exercise induces a certain degree of bone atrophy.

Thus the calcium reservoir in the bones is smaller than in penned fowls. When laying begins this calcium reservoir is strained to the limit.

But why does such a bird not stop laying, or at least produce eggs with poor shell quality?' asked Dr Siller. This last factor was unexplained.

H. R. C. Kennedy of *Farmer and Stockbreeder*, Advisory Service, 30th January 1962, comments:

*Fatigue can be regarded as synonymous with exhaustion and is a condition which can account for a considerable wastage in some flocks. In laying batteries, where individual egg production can be recorded, outstanding layers can be found dead, the day after laying, with no preliminary ailing period or indication of disease, nor indeed does a subsequent post-mortem indicate the presence of any disease or even reproductive disorders – just death from utter exhaustion.*

*Such losses seem to occur more frequently in laying batteries and all-wire or slatted-floor systems, than where the layers have access to deep litter, and this could appear to indicate the possibility of a nutritional factor being involved.*

*The modern layer is, after all, only a very efficient converting machine, changing the raw materials – feedingstuffs – into the finished product – the egg – less, of course, maintenance requirements.* (The italics are mine.)

White Leghorns have always been nervous and highly strung and they form the basis of today's commercial flocks:

Closed breeding flocks of necessity involve some degree of inbreeding and this not only emphasises good characteristics, but can do the same to undesirable ones; some of the latter may be ultimately eliminated, while others cannot.

Another danger of intensive inbreeding of flocks for a specific purpose is raised by a correspondent to *Poultry World*, 18th October 1962:

*Provided we can give the birds all they need, apart from the good earth, there is no reason, as I see it, why intensive breeding should not be a success. The one danger in doing this is that we may breed birds that can only do well under intensive conditions. That will not matter if all the progeny are kept intensively.*

There is also the difference between laying battery and deep litter or 'flock' conditions. Here I use the word 'battery' to mean not more than two birds to a cage. All other intensive systems can be called 'flock' conditions.

*It is common knowledge that birds which do badly under flock conditions may do well in a battery. This condition is likely to be inherited; if stock is selected from laying battery records, we may produce strains that do well only under battery conditions. If we select birds under intensive flock conditions we may breed strains which will only do well under these conditions.*

*It is likely to take a number of generations of such breeding to bring this about; but it could show worsened results in only a few years.* (The italics are mine.)

Mr Kennedy wonders if diet is not a stress factor for the laying bird:

No breeder would expect to get good hatchability or to produce strong chicken from his breeding stock, unless that stock was fed on a breeders' quality diet, yet commercial layers are expected to maintain high levels of production on diets that would not be suitable for breeding stock whose production is lower.

These foods do not, however, contain the range of amino acid proteins or of the B group of vitamins contained in the breeder quality foods considered necessary for the production of hatching eggs and supplying the growth requirements of the embryos that will develop.

The Ministry of Agriculture in their leaflet *Incubation and Hatchery Practice* also point out that the diet adequate for laying birds is quite inadequate for breeding birds and does not produce good enough eggs for hatching:

The fact that commercial laying hens will produce well on a variety of laying diets should not be allowed to give the impression that the same diets are adequate for breeding stock. Emphatically this is not so, and the use of such diets will soon lead to troubles. Furthermore, the fact that breeding stock are laying eggs is no guarantee that even if these eggs are fertile they will hatch, or, if they do hatch, that the chicks will be able to develop properly. The usual limiting factors are deficiencies of vitamins or minerals in the egg, and it must be realised that the degree of deficiency that is enough to render hatchability a matter of speculation need not have any ill-effect on the health or the productive performance of the hen.

The Ministry then goes on to describe the possible results of deficiencies on the chicks when hatched from laying eggs:

Deficiency of animal protein factor, principally due to deficiency of vitamin $B_{12}$, leads to a rapid decrease in hatchability with progressively poorer survival of those chicks that do hatch. Deficiency of riboflavin leads to poor hatchability, with a high incidence of malformed embryos showing oedema and clubbed down. Pantothenic acid deficiency lowers hatchability, with a high incidence of apparently normal embryos dying over the last 2 or 3 days of incubation. Biotin, choline and manganese are required for normal development of the embryo and the prevention of the condition known as enlarged hock, slipped tendon, or perosis. Acute deficiency of biotin causes high embryo mortality over the period from the 72nd to the 96th hour of incubation. Choline deficiency alone is unlikely to occur on British diets as the hen seems fully able to synthesise her requirements.

A. C. Moore, in *Poultry World*, 22nd November 1962, was consulted to see if he could explain why flocks of pullets suddenly and for no apparent reason dropped from 70 to 20 per cent production, and he commented that:

> *… pullets … geared almost to breaking point, become victims of stress factors.*
> *The more flocks of pullets I examine the more convinced I become that productivity is outpacing vitality.* And now that an attempt is being made to maintain numbers of eggs with increased size, these mysterious problems will become more numerous.
>
> Recently I was on a farm where intensively kept pullets were subjected to a continuous blaring from wireless loud-speakers. These had been installed, not because it was thought that the music had charms which stimulated production, but to provide a continuous noise.
>
> Feeding and watering were automatic, cleaning out was done annually when the houses were empty and therefore these flocks were entirely on their own, except for a brief period daily. Any outside voice or a sudden entry into the house could cause nervousness which affected production.
>
> *It is difficult to avoid the view that poultry stocks today are suffering from nervous tension, due to our attempts to convert them into egg machines.* (The italics are mine.)

Automatic feeding has another hazard for the bird: in his book *Hen Batteries* Dr Blount warns that even with automatic feeding devices a careful watch must be kept to see that they are working properly and the troughs kept well filled. Otherwise, he says, ' *it is possible for the birds to develop a* frustration complex *by which they* fail *to lower their heads to the trough, and semi-panic when they do not get any food.* They paw at the cage front and instead of lowering their heads to the feed they raise them higher and higher!' (The italics are mine.)

Cage Layer Fatigue is not the only disease causing worry to battery owners. Apart from fowl pest many other diseases have increased alarmingly during the past ten years. Mortality in the battery house averages at 12 per cent to 15 per cent, and including birds removed from the flock before they have actually died, can be well over 20 per cent. Twenty out of every hundred is a lot of birds! It appears that digestive diseases account for some 15 per cent of deaths, and half these are maladies of the liver. Fatty degeneration is a disease typical to battery birds, who are given plenty of food and get overfat. About one bird in six dies from some disorder in the reproductive system. Diseases of the excretory system have increased tremendously and nephritis has sometimes reached epidemic pitch, when it is called Pullet Disease. All forms of cancer form a large proportion of mortality cases and Dr Blount records that in one year alone an experimental unit recorded '*cancer* of the heart, lungs, ovary, oviduct, kidney, leg muscles, liver and abdomen'. He thinks there is a correlation between the increase in egg production and in some aspects of the Avian Leucosis complex, and in diseases of the reproductive system. He also found a correlation between the increase in egg peritonitis and the extra night lighting the birds get.

Disease control these days is remedial rather than aimed at exterminating the root causes. This is possibly because the birds are now only expected to lay for a year and are then replaced by new birds, whereas formerly birds were kept for a second or even a third year. The moulting period of six weeks at the end of a year's laying is, after all, uneconomic, as the bird is not then laying eggs and paying for its keep.

Dr Hutt, Ph.D., D.Sc., Professor of Animal Genetics in the Department of Poultry Husbandry in Cornell University, U.S.A., gave the impression in a lecture on mortality in poultry that:

> ... the American system of poultry keeping consists largely of vaccination and drug application.
> On this matter Dr Hutt indicated the availability of controls of this sort had resulted in a deterioration of management practices. He expressed the opinion that resort to drugs should be made only when sound management and good practice had failed to eradicate a disease incidence. (*Poultry World*, 12th October 1961)

One feature of intensive poultry keeping not yet mentioned is the use of insecticides to keep down flies and parasites. I shall have more to say on insecticides and drugs in a later chapter and it will be sufficient here to quote a reference to research undertaken to test the effect of systemic insecticides (*Farmer and Stockbreeder*, 29th August 1961):

> Now what do you think happens if you feed selected insecticides continuously? One trial in recent months has done just this and (hardly surprisingly) toxicity was expressed as mortality, lower egg production, poorer growth or weight loss. Birds were fed this way for 29 weeks. Curiously enough, some lived to tell a tale.

Increased egg production brings headaches as far as egg quality is concerned. Poor shells, thin shells, pale yolks, watery whites, all prove a problem. For poor shells and thin shells improvement can be made with additions to diet, while emulation of the golden yolked free range egg can be got by feeding dried grass or, if that prove too expensive, by feeding a yellow dye to the bird. In Australia, producers receive a premium for golden yolks, and this is beginning to operate in some packing stations here. The housewife, curious creature that she is, feels the egg she used to know had more quality than its pale yolked successor. She sticks to this idea stubbornly, against all assurances to the contrary, and is willing to pay more for her free range eggs, so the free range egg must be simulated as far as possible. The bird can only co-operate in this endeavour as far as she is able. 'Severe de-beaking can lead to poor shell quality by restricting the birds' ability to eat calcium grit' comments a report in *Farming Express*, 31st January 1963.

A. C. Moore, a correspondent to *Poultry World*, 22nd November 1962, wrote

> It is time all concerned with the interest of the poultry industry and the sale of eggs to the housewife began to pay some attention to the decline in the size of egg yolks in relation to the increase in the amount of white.
>
> I am often humiliated when my wife calls my attention to the smallness of the yolks when breaking eggs out to fry or pushes over the boiled egg she is trying to eat for me to examine.
>
> Some years ago it was hardly possible to cut off the top of an egg without touching the yolk. Now you are more than a quarter of the way down the egg before the yolk is reached.
>
> Some weeks ago I was preparing a salad supper which involved the boiling of several dozen eggs. When these were cut up, the yolks of some were no bigger than walnuts.
>
> When I was a member of the No. 3 B.E.M.B. (British Egg Marketing Board) Regional Committee, I pointed out this decline in yolk size to the technical advisers of the Board and asked if some investigations could be carried out to ascertain the exact weight of the yolk in relation to the varying sizes of eggs. I have never seen a report upon any such investigations.
>
> I am told by housewives that *they now need three eggs to provide the same amount of yolk which two once supplied....*
>
> ... I suggest that the increased production of present-day hens causes the yolk to be shed from the ovary before it is developed. Neither does the speed of passing down the oviduct help to improve size. (The italics are mine.)

But 'What's Wrong with Egg Quality?' asks *Poultry World*, 4th October 1962:

> Confessing to being alarmed at the prospect of too much publicity being given to the subject of internal egg quality, Mr Cyril Thornber commented at a Kent Egg Producers' Association meeting at Bearsted last week: 'I hope we are not going mad on this topic like they have in America.'
>
> The longer the Americans worked on the subject the less progress they would make on other aspects of egg production, he stated.
>
> It was a severe restriction on the breeder because eggs with the best quality albumen had the worst hatchability. 'The Americans will probably end up with excellent egg quality but with an increase of around 15 per cent in chick prices,' he said.
>
> Internal quality of British eggs was quite satisfactory in his view and the best way to improve it further was to get eggs into shops more quickly.

American scientist Dr Hutt did not think extra quality worth the extra cost (*Poultry World*, 5th October 1961):

> In the first of two talks in this country, Dr E. B. Hutt, Ph.D, D.Sc. (Edin.), professor of Animal Genetics at Cornell University, discussed the question of egg quality from the point of view of the breeder.

Could the breeder do anything about improving quality? he asked. 'Yes, he can, but it comes down to the question of whether it will pay him to do so,' declared Dr Hutt....

Turning to interior quality, Dr Hutt indicated that this was not so much a breeder's problem. Albumen quality would decline if eggs were not stored properly, although it was true to say that those eggs which were best to start with declined in quality the slowest. But there was some correlation between the rate of laying and egg quality.

One trouble was that the egg marketing specialists had emphasised too much the question of albumen quality. 'But if the housewife is satisfied with a lower albumen quality why try to dissuade her and so create problems for yourself?' he asked.

Yolk colour did not depend on the breeder but on the food. One could change the chemical composition of an egg by putting certain substances in the diet but it did not necessarily pay to do so. One could feed to produce eggs with higher vitamin A and D, for instance, but the producer would not get any more money for a super-vitaminised product.

One producer confessed (*Poultry World*, 14th June 1962):

'I am in this business for what I can make out of it. If it pays to do this or that, I do it and so far as I am concerned that is all there is to say about it.

'If my eggs are not as good-looking as they might be, that is no concern of mine, I pay the Egg Board to sell them, so why should I worry.

'I am not concerned either about the quality or appearance of my eggs. My worry is getting my eggs over the grading machines.' ...

Leaving the place, a small pen of Rhodes was noticed to be installed in a secluded corner. It was intimated that they were kept to 'supply the house'!

This last paragraph is reminiscent of the broiler growers who keep chickens in their back yards 'to supply the house'. Perhaps that is why the 'foolish' housewife travels far and wide in her search for quality eggs.

Consumption of eggs has been rising steadily in numbers to cope with increased supplies. In 1953, when feeding stuffs were derationed, it was estimated that we ate 200 eggs per year per head of population, in 1957 222 eggs, in 1959 240 eggs and in 1960 250 eggs. Advertising gimmicks aim to persuade us into eating still more eggs. 'Go to work on an egg', 'Start the day on an egg' are familiar slogans. A well-known series of advertisements on television featured a mock traditional country farmer complete in his shabby clothes and with an engagingly broad accent, quite unreal to this age of commercial egg production. The early advertisements in this series showed him collecting his eggs from under hedgerows. They were aimed clearly at the urban housewife. Perhaps the housewife was not quite as gullible as was at first thought, for the delightfully fictional touch was soon dropped and replaced by a sunny outdoor farm scene, it being hoped, no doubt, to convey by this more modest subterfuge an impression of quality and goodness.

*Farmer and Stockbreeder* (perhaps ironically) advises small producers who rely on sales from the gate:

> From the many wire and slatted floored small-scale conversions up and down the country those with a hundred or so birds are going to look smugly on the various integrations and take-over battles which will rage in days to come. Not for them the rat-race of modern society. Instead most probably they will operate quite as efficiently with an 'Eggs for Sale' sign at their front door, forget about gradings and be content to let the hen put her own trade mark on eggs sure to sell at a premium....
>
> It doesn't pay to be too lackadaisical in the approach to direct sales. Granted at this moment you can't go wrong in putting up a pretty sign and mentioning casually that your birds are on free range. Select the eggs in the customer's view from a big basket, with a few wisps of straw and a feather or two floating around, and you can claim sixpence premium right away.
>
> That might not last. There is just a chance that too many will employ this safe gimmick. But if you transfer the eggs from the basket to some sort of attractive container then you win. (11th April 1961)

But Mr W. G. R. Weeks, agricultural economist of the University of Durham, pointed out that American tactics must be used to cope with increased output (*Farming Express*, 21st December 1961):

> Value of eggs to health – bringing beauty to girls and building boys' muscles – are themes to exploit when selling eggs for teenagers.
>
> And egg cartons illustrated with nursery rhymes and cartoon characters are already being used in the United States to appeal to children.
>
> These are so successful that packers are having difficulty in finding enough small eggs to sell to young people.
>
> Marketing idea borrowed from breakfast food manufacturers is a series of plastic toys in a branded egg pack.

And in *Farmer's Weekly*, 1st December 1961, he wrote:

> The U.S. Poultry and Egg National Board do useful work in making the American homemaker egg-conscious. Typical examples of their publicity are 'All this and only 77 calories!' featuring two gorgeous poached eggs and bacon strips. 'Eggs. Their remarkable value in infant and child nutrition' – A beautiful mum and cheery child tucking into two boiled eggs apiece. 'You are what you eat' – this last slogan featuring a sylph-like housewife enjoying a three-egg omelette.

Stimulighting and twilighting are not the only pieces of research to gladden the hearts of profit-seeking poultrymen. 'Birds Without Combs Lay More – Pay More' stated an article in *Farming Express* (18th May 1961), 'Pullets with their combs removed lay more eggs, eat less food and thus show a bigger profit....'

> Two other factors were noticed. Dubbed birds had the biggest advantage over controls in winter. The liquid intake of the controls declined in cold weather.

The two factors were related. Birds did not like dipping their wattles in cold water and their liquid intake in winter was therefore reduced.

On the other hand, in temperatures of more than 80 deg. F. the dubbed birds suffered most. Loss of the combs and wattles removed part of the heat-loss mechanism.

Other advantages claimed for dubbing were increased docility because the birds could see better without the floppy comb over their eyes.

Dubbed birds did not injure so easily, particularly in wire cages.

Several large hatcheries in the United States are offering dubbed day-olds at an extra charge. The operation is carried out with curved nail scissors.

Even with the best management some birds will not settle down 'two in a bed'. This can be got over by putting hen specs on these two birds. By the time they get these off they will have settled down all right. (*Poultry World*, 27th December 1962)

We have seen that the chief aim of intensive egg producers is to make the chicken into a super-efficient machine for laying more and more eggs in a given time, and if, after all, she bears little relation to the chicken as we knew it, who cares?

'Frightening Things Happening' read a headline in the *Farmer and Stockbreeder* (23rd October 1962):

Attempts to increase egg production from hens by crossing them with the Japanese quail are being made in Britain, revealed Dr Rupert Coles, Ministry chief poultry adviser, at a B.O.C.M. conference in London. The quail produced a large number of small eggs with a very low feed consumption and it was hoped to combine these features with the hen's larger egg size.

Certain problems had to be overcome, said Dr Coles. These included the sterility of the first cross of any hybrid, but he thought there were possibilities.

Dr Coles was commenting on what he described as 'frightening developments' in poultry breeding which had been forecast at the World's Poultry Congress in Sydney. Irradiation of day-old chicks to completely change their characteristics had been mentioned. With this technique it might be possible to produce any type of laying hen merely by the application of special rays to the chick, whatever its parentage.

Another possibility was the further development of parthenogenesis – reproduction without sexual union – and male birds might become unnecessary. There was talk, too, of giving hormone injections to hens so that an egg was laid every six hours and production totalled about 350 eggs in six months.

'These are not fantasies,' said Dr Coles. 'Enough scientists spoke at the Congress to show that all are live projects.'

In September 1962 appeared eggs in plastic containers aimed to show the housewife how fresh they were, and also to allow her to cook them in the container. Resistance to breakage was another selling point.

But despite gimmicks does the Egg Board manage to sell all the eggs or are we over-producing? On 28th June 1962 *Farming Express* reported:

> The Egg Board has millions of pounds worth of liquid egg in cold store. This represents from 15,000,000 to 20,000,000 dozen unsold shell eggs. And the quantity is growing.
>
> Last year Britain imported 16,662,000 dozen eggs from Poland. For months the board has drawn thousands of dozens of eggs off the market each week for breaking out.
>
> Home production has been enough to meet entire market demands for shell eggs.
>
> And to prevent prices slumping in a surplus, the board has been strengthening the shell egg market by breaking out and freezing some of them.
>
> In April last year the board had £2,000,000 worth of frozen eggs in store. This year, according to a board official, stocks are higher than normal.

And on 6th November 1962, *Farmer and Stockbreeder* warned:

> Dramatic upswing in the egg supply position with a serious risk of over-production is anticipated in the poultry industry. Throughput is running at 13 per cent above last year, with shell egg sales only $6\frac{1}{2}$ per cent higher and the remainder being processed. In many respects the situation is similar to that of 1959 when the British Egg Marketing Board Chairman, Mr W. J. Welford, made a personal appeal by letter to producers to cut back on output. This much criticised action is likely to be repeated shortly….
>
> Although recording the fact that severe competition from imported frozen egg affected prices, the accounts show that processed egg stocks at the end of March were similar to 1961. But Board officials have not revealed the CURRENT level of stocks which are believed to be embarrassingly high.

Despite the air of smug efficiency which hangs over the intensive poultry producer these days, subsidies are still being paid, by the taxpayer, for eggs. The flat rate of subsidy is the difference between the guaranteed price per dozen fixed at the year's Price Review and the amount at which the Egg Board reckon they can sell the eggs. At the beginning of 1961–2 the guaranteed price was 3s. 8·63d., the estimated selling price, 3s. 3·2d. and the subsidy 5·43d. By March 1962, however, increase in poultry food prices caused an increase in subsidy to 8·09d. This had dropped again by June to 7·49d. per dozen eggs. Total subsidy for the year to 31st March 1962 was £20·9 million.

In 1958 producers received 4s. 6d. a dozen large eggs down to 2s. 8d. for small eggs, in 1961 they received 3s. 9d. for large eggs and only 1s. 7d. for small eggs, another hazard of over-production.

And what happens to these layers at the end of their year of hard production? Their lack of exercise means that they remain well fleshed and soft, so that an attempt is made to sell them as table birds. The Plant Committee commented that 'specialised lightweight laying hens are of less value as table birds. Many of these birds go to food manufacturers who use them to prepare poultry meat products.'

It was a great problem to producers to know what to do with these light hybrid layers at the end of their laying life, and many found the return on their carcases too small to bother marketing and they were burned or buried. They were too light to be sold as roasters and too old to be handled by broiler plants. But in the summer of 1962 two packing stations began to process them like broilers, one called them 'baby boilers'. Another suggested use for them was as pet food. It was reported in August 1962, however, that these hens were only bringing a return of 6d. a pound to producers, and in *Poultry World*, 31st January 1963, the gravity of this small return to the producer was stressed:

> Before the war displaced laying birds in many cases paid fully for their replacements but today there was a gap of something like 10s.
>
> Emphasising the gravity of the situation that had developed so gradually over the years as to have almost escaped notice, he declared that an egg producer relying on 1,000 birds to give him a weekly income of £10, had also to budget for a loss of £10 a week to meet his replacement obligations.
>
> That meant that if a man with 1,000 layers was not making £20 a week out of them he was having to eat into his capital to live.

It may well be asked whether this rat race is worth while? It has had the effect of flooding the market with more, cheaper, smaller and inferior eggs, but is this really what the public wants? and does it serve the best interests of the community?

# Veal Calves

An unavoidable characteristic of the rearing of animals is that approximately the same number of male and female offspring will be produced. It follows that where cows are kept for milk there is the problem of what to do with male calves, many of which are not suitable for rearing as beef because the strain has been developed primarily for its milking potential. It has been estimated that the surplus of unwanted calves, 'bobby' calves, in this country amounts to some 800,000 to 1,000,000 yearly.

What happens to them?

In the first place there is a high mortality rate.

> Each year hundreds of thousands of calves die before they are a few weeks old (*Farming Express*, 7th September 1961). If they could be saved for rearing, beef imports could be cut by up to £50,000,000 a year. Our marketing methods which subject young calves to their greatest abuse when they are least able to withstand it are to blame for this wastage.

The reason behind this wastage is manifestly that these bobby calves are as much an encumbrance as an economic asset.

The harsh conditions to which calves are subjected in transit and in markets represents a hazard even for the calves which are bought for rearing. Many farmers, by taking for granted this wastage and sending their calves to market without any food in them, nullify their chances of being bought for any other purpose than slaughter.

> Many very young calves are so weak when exposed for sale in the
> markets that the observer wonders how little, if any, colostrum they have
> been permitted to have before leaving their farms,

wrote a veterinary surgeon in *U.F.A.W. Courier*, Autumn 1960.

The bobby calf is separated from its mother at birth or a few days
after. It is taken, often without a feed inside it, and bundled into the back
of a truck, exposed to the cold and rigour of a market, to the cruelty of
some drovers with their hobnailed boots and sticks. Neither the drovers,
nor the boys who help them, appear to notice the distress of these young
animals, indeed the children follow the men's example of whacking the
sides of pens where animals are quietly lying momentarily oblivious of
their plight, laughing as they start up again in fear.

After the rigours of the market some calves travel hundreds of miles
to veal centres in cold, overpacked lorries, to await their turn at the slaughterhouse. Slaughterhouses are not compelled to feed an animal unless they
keep it for more than twelve hours.

So these gentle little creatures meet their end, a few days after
being born, and having experienced nothing but hunger and fear at our
hands. Their carcases are sold for a pound or two to the manufacturing
veal trade, who use them in pies, tinned foods, cutlets and pastes, and
to the leather trade for whom their soft skins can fetch a premium in
gloves and shoes.

Some farmers buy and collect calves personally so that they shall not
arrive at the farms too weakened by cold and rough handling. Calf banks
are being started in many districts, so that farmers may buy direct from
other farms and know that their calves are not only going to avoid the
upsets of the markets but also the risk of infection due to decreased resistance caused by exposure and fear.

To these, and similar better tended stock, especially the lighter dairy
bred calves, the Friesian, Ayrshires and Channel Islands, the 'quality veal'
trade now gives a three months respite from death. In this industry, as in
all intensive farming, producers are warned that, although they choose
from the bobby calves as opposed to the more expensive rearing calves,
they must choose bright-eyed, healthy and vigorous calves, bouncing with
vitality. Weaker ones would not stand up to the life.

A limited quantity of veal has been produced in this country for hundreds of years. Until about a generation ago it was the practice to allow
calves to suckle for about six weeks before slaughter. Even in those days
it would appear that there was a call for very pale fleshed veal because it
was the practice, once a fortnight during the animal's life, to nick a vein in
its neck and allow the blood to drain away.

> … It is not so very long ago since our own Essex veal men were
> administering black pepper to calves to make them drink more milk, and
> bleeding them fortnightly to keep the flesh white. (*Farmer and Stockbreeder*,
> 13th September 1960)

This useless blood letting was the farmer's primitive attempt to provide the public with what it demanded – or what he supposed it demanded – and all the recent developments in the veal trade have similarly arisen as an attempt to provide a yet whiter and whiter flesh.

I shall discuss to what extent this is or is not a desirable goal in another chapter, here I deal with the techniques now used to produce this super-white flesh.

On the continent, and especially in Holland, white veal has been produced by specialised methods for over a hundred years. Calves were fed exclusively on whole or skimmed milk and many thousands produced annually. Losses were severe because this was a deficiency diet the control of which was not sufficiently understood.

During the last ten years, however, milk substitutes have been patented in Holland which reduce the cost of veal calf rearing to a third of the cost when whole milk was used, and which, by the addition of some minerals and vitamins also reduce the incidence of calves dying before they reach the slaughterer. This, for the first time, has made veal rearing an economic proposition and created the veal industry as we understand it today.

Before this time the veal trade in Holland, as in England, was beset by curious and improbable methods. Housing conditions for the veal calf could be appalling. The calf was frequently

> … placed in a very small pen surrounded by straw, usually in a kind of box where they were more or less packed tightly with straw all around them (wrote Dr Bakker in *The Veterinary Record*, 27th August 1960). The pens were always in a dark corner, and to increase the darkness a lid was very often put on top of the box with a few holes in it for the calves to breathe. The theory was that immobility increased the rate of growth and that darkness favoured the production of white flesh – probably because plants reared in darkness are white….

Or in another reference, this time speaking of more organised houses:

> Formerly all these Dutch veal houses were kept in darkness, with lights switched on only at feeding time. Some are still darkened, and it is a peculiarly pathetic sight to see, when the lights are switched on, a hundred calves struggle to their feet – if they have been able to lie down – and push their heads through the feeding holes for the buckets which are their only diversion from – just breathing. (*Farmer and Stockbreeder*, 13th September 1960)

Bloodletting in England, darkness in Holland; both products of ignorance and an attempt to please a gullible public. But the Dutch are a persevering race and by trial and error they have managed to build up a vast industry. Holland now rears about four hundred thousand veal calves a year, keeping a few for the home market and selling mainly to France, Italy, Germany and the U.K., to whom she sends some six to seven hundred carcases a week. Our own veal calf industry has been built up entirely in simulation of Dutch methods, both good and bad.

At first Dutch milk substitutes were imported by British veal farmers, but comparable products have been developed here which also form the sole diet of the calf until it is slaughtered at three months. These home-produced carcases are marketed as Dutch ones, I was informed, because English veal is still associated with bobby calves which, naturally, have very little flesh on them. We now produce around twenty thousand veal calves a year, spread over some forty farms of varying holdings.

Like the Dutch, we have plunged into yet another form of animal industrialisation without knowing very much about it.

> Calf rearing is about the last of the farming processes to become industrialised (commented the Editor of *Farmer and Stockbreeder*, 13th September 1960). It has, until recently, been a predominantly domestic occupation, or a cowman's spare-time job. Now the calf goes into some sort of farm factory, like the cow, the baconer and the hen.
>
> The broiler calf for veal or early beef had to come sooner or later. But the movement is now revealing how little we know about the calf.
>
> What are the comparative growth rates of vealers? What diets make offal and what make the more valuable flesh? At what temperatures should the broiler calf be kept in its milk and early ruminating periods? Can hybrid vigour make any contribution to profits?
>
> The questions flow on and on. Some are partially answerable. Some are quite without an answer yet.

The English producer uses only bobby calves and aims to produce in twelve to fourteen weeks a white-fleshed calf of 220–280 lb. liveweight and 140–170 lb. deadweight, and all effort is directed to this end. The Dutch choose a red and white Friesian calf, or a black and white, which costs a lot more money initially, upwards of £15, but which grows very rapidly and achieves a much greater weight than those we produce.

> I use the word (veal) 'manufacture' deliberately (said R. Trow-Smith in *Farmer and Stockbreeder*, 13th September 1960), because this is a factory process. For it, you need no land at all. As one of the Dutch calf food makers said to me, 'you can start in your back kitchen'.

In a way this is the impression I first got when I began to investigate this trade. Buildings for housing veal calves are extremely varied, from scien-tifically built and insulated environment controlled buildings designed for the job, or big airy barns, down to sloping corrugated iron lean-to's. And the inside shows equal variation according to the owner's ideas of what produces good veal.

At the worst places I have visited the calves were kept in solid-sided crates no more than twenty-two inches wide by five feet deep. This is just big enough to house the calf standing, but barely enough to allow it to lie down except in a very restricted manner. The added fact that it is housed on slats makes any movement in the confined space still more uncomfortable.

It was kept in the dark apart from two short periods a day when a shutter is let down at the front of the crate for feeding. When we entered the building and the farmer switched on the light there was pandemonium from within the crates. He had to talk to a calf soothingly for many seconds before he dared to let down the shutter, and then there was no mistaking the misery on the face of this calf with its enormous staring eyes. The farmer himself seemed abashed and unhappy at the sight of it, but reassured himself with the remark that it produced top quality veal. The photograph, Figure 1, was taken when the calf had been in the light for some time and had calmed down. It is only fair to say that this was an experimental unit for the purpose of trying out the Dutch system and in many other respects this farm was a model of its kind and showed that there was warmth and care for the stock.

I do not think that there are many holdings in this country where such extreme conditions prevail. It is instructive to notice that this same firm write in their pamphlet on veal production: 'The idea that calves must be kept in crates and in total darkness is quite erroneous....'

Another farm had the calves standing in a row on a slatted platform, their heads held between two vertical wooden bars so that they could slide up and down and nothing else. They could slide down to a lying position, but their necks would still be relentlessly held, they would have only that one position of rest throughout their lives. It reminded me forcibly of a row of stocks. These calves were indescribably dirty and were obviously suffering badly from the flies milling around them. They were shying up with their back legs, but of course could do little to alleviate their misery.

For the calf 'soiling can ... be a source of serious discomfort' (*The Veterinary Record*, 27th August 1960). A healthy calf can lift up its tail and avoids dirtying itself, but these calves have less energy and some do get very dirty and can do nothing to get themselves clean.

On the other hand, on another demonstration farm the calves were housed in seemingly perfect conditions. They were in pens of roughly twenty square feet, bedded comfortably on straw, in a barn with large openings allowing the sun to stream in. They were housed singly or in pairs, the pairs having forty square feet, and their housing conditions differed in no obvious way from those of the adjoining beef calves. When fresh straw was put down it was sprayed with Jeyes' fluid to prevent the calves browsing, but they looked so much more comfortable than those tethered on slats. Unfortunately there seem to be very few farms where calves are reared in such comfort. Curiously again this firm advocates in its booklet: 'veal calves should never have direct access to hay or straw because the action of rumination will tend to reduce the killing out percentage ...' and again, 'direct sunlight should certainly be excluded for this will tend to make them restless'. Despite these warnings their calves have won many veal carcase competitions.

These, then, are the worst and the best conditions I have seen. But what are the standard types of housing recommended to producers?

**1.** Communal pens in which there is just sufficient room for all the calves to lie down at once. In these the calves are sometimes tethered round the edges.
**2.** A slatted standing with vertical bars along the front to which the calves are close tethered.
**3.** Individual pens with slatted sides, twenty-two inches wide and four to five feet deep, the calves yoked to the front.

In each case slats are advocated as being the most practical from the cleaning point of view. However, the two farmers I met who used straw to bed down their calves, did not clean out the pens except between batches of calves, fresh straw being added regularly on top of the old. They did not find that the labour involved at the end of a batch was any greater than if they had used slats. Calves up to two weeks of age are, in any case, allowed some straw for warmth and comfort, but after that age they are old enough to nibble it and so it is taken away. The reasons for this I will explain later in this chapter.

Subdued light and an even temperature of 60° to 65° are also recommended, and 70–75 per cent humidity should not be exceeded.

The most easily managed type of housing, and consequently that most widely used, is the individual twenty-two-inch pen, with slatted sides so that the calf can see its neighbours, open at the back to facilitate cleaning, and with a holder for its milk bucket in front. The calf wears a collar which is close-tethered by rings to posts at the front of its pen permitting it to stand or slide to a lying position, but not allowing it to turn round or lick itself. One farmer told me that his calves nearly went mad the first day they were tethered and then seemed to resign themselves to it. On a small-holding there is no room for differences in temperament of calves, but on a larger mixed farm there is more opportunity to take out the calf which cannot tolerate the close confinement. 'I put them in for baby beef,' one farmer told me pointing to the covered yard where the baby beef calves could roam freely on the straw. On this farm vealers of a contrary disposition were obviously the lucky ones.

Some veal producers in this country set out to introduce methods here which, they hoped, would eliminate some of the worst aspects of Dutch rearing. Whilst keeping to the latest ideas of ventilation and temperature, they housed the calves in pens holding two or more calves, allowing ten to twelve square feet per calf, and keeping them untethered. Not unlike the average herd replacement conditions except that calves reared normally have more than twice the space allotted the veal calf and have the warmth and comfort of straw as opposed to slats. They started with good intentions but, except at one farm I visited, the intentions fell through because the calves could not be cured of suckling and urine licking. The British Veterinary Association booklet *The Husbandry and Diseases of Calves* states:

> Almost all calves which are given a liquid ration with little access to roughage develop chewing and licking habits; those affected with lice are particularly prone to the latter. It is common to find hair balls varying in size from ½ to 2 inches in diameter in the digestive tract.

I was shown a hair ball the size of a cricket ball taken by a slaughterer from the carcase of a veal calf. '... Preventive measures involve the provision of an adequate diet and the early inclusion of solid food in the ration. Regular grooming also helps.' These ameliorative measures, in the context of their trade, are not open to the veal farmer, and so great is the calf's natural craving to suckle that the only other remedy open to a veal farmer, namely that of tethering the calf for an hour after each feed, also failed. So they had to return to the methods they had tried to eliminate, that of individual penning and permanent close tethering.

The dual aims of veal production are firstly, to produce a calf of the greatest weight in the shortest possible time, and secondly, to keep its flesh as pale as possible to fulfil a real or a supposed consumer requirement.

Rapid weight gains are achieved through immobility; thus food energy all goes into weight gain and none is lost in frisking or exercise of any sort; through high level milk feeding, the veal calf receives on an average four gallons a day at twelve weeks, whereas other calves get only two gallons; and through a far higher proportion of fat in their milk than is normal. The veal milk substitute has, as we shall see, about 18 to 20 per cent fat as compared with the 1 to 2 per cent in other milk substitutes. The conversion ratio of milk as opposed to solids as a diet is far higher and putting on weight involves in the calf the indistinguishable processes of growth and fattening. Dr J. H. B. Roy of the National Institute for Research in Dairying, says:

> In estimating the energy equivalent of weight gain in young cattle, an immediate difficulty is that high weight gains are associated with fattening and the demarcation of growth and fattening is not clear-cut. Growth involves the deposition of protein, ash and water with little fat, whereas fattening involves deposition of fat with little protein, but both can occur together in the young growing animal. (*Scientific Principles of Feeding Farm Livestock*)

High level milk feeding introduces its own particular problems. The powdered milk substitute is reconstituted with a large amount of water to simulate milk and make it palatable. Therefore in inducing the calf to consume the maximum amount of milk substitute it must be persuaded to take in more liquid than it needs to sustain it in its static and immobile state.

The manner in which this is achieved is ingenious.

Between feeds it is allowed no water at all. The calf is kept in a steady temperature of around 65° – or more if there is a heatwave. It sweats, and this loss of moisture makes it thirsty. At its next feed it drinks to excess, sweats again and again becomes thirsty, and so on.

This is explained by the following quotation from *Farmer and Stockbreeder*, 13th September 1960:

> The Dutchman likes to see his calves sweat, not from high external temperature but like an executive after lunch, from rather too much to eat too often.

And J. H. B. Roy wrote in the same issue:

> There is no doubt that even now there is unanimous opinion in Holland that veal calves should sweat, and calves that do not are often rejected as bad doers. It is known that at any given level of milk intake, the higher the environmental temperature the greater the tendency for the calves to sweat, and presumably at any given temperature, the greater the quantity of milk given the greater the inclination to sweat.
>
> As sweating is used by the Dutch as a criterion that a calf is being fed to its maximum potential, the best results might be expected from calves fed to appetite at a particular weight, in temperatures which had been raised until the animals were just sweating. Increasing the temperature may also have a direct effect in increasing the thirst and therefore the appetite of the calves.

In the Ministry booklet, *Calf Rearing*, T. R. Preston of the Rowett Research Institute wrote: '... by the time the calf is ready for slaughter, at about 12 weeks old, it is consuming from 4–5 gallons daily. To get the calf to drink these large amounts no water must be given.' The success of the operation becomes apparent when one considers that a calf of this age would only be taking about three gallons of water in toto.

Weight gain, though it has its own complexities, is a far easier problem to determine than that of keeping the flesh white, which is a highly intricate and scientific process deserving discussion in some detail. It involves two interrelated processes, firstly preventing the development of the colour pigment myoglobin in the muscles or flesh of the animal, and secondly giving the calf a diet designed to produce a degree of anaemia. All young animals are born with pale flesh and this rapidly darkens with exercise, food and age.

As I have shown, the veal calf is confined so closely that its only movement on the average farm is by standing up and lying down on its tether. There seem to be various reasons why movement is not allowed. A veterinary surgeon, writing on veal calf production in *The U.F.A.W. Courier*, says:

> Management is directed towards converting every available ounce of liquid fed into carcase weight and finish in the least possible time. In order to favour rapid food conversion and *also to prevent the development of muscle pigment in the flesh*, movement is strictly limited by keeping the calves in close confinement. (Autumn 1960. The italics are mine.)

The addition of tethering is to prevent the two 'vices' of suckling and urine licking.

Suckling is a craving shared by all calves which are separated too early from their mothers. Other calves can be given hay to chew after their milk feed, or at most can be tethered for an hour after it, and the trouble is mostly overcome, but these veal calves must not have roughage so they are put in individual pens where they can have no contact with other calves and suckling is impossible.

Urine licking is a diet deficiency habit due to insufficient iron in their milk. Calves have a marked sense of natural cleanliness, like pigs, and do not normally go anywhere near their urine.

> Any trace of iron in what the calf eats will cause discoloration. Straw contains a little iron, and, as the calves long for iron, straw is barred and slats provided instead. But so desperate for iron will the calf become that it will lick the slats that are impregnated with urine, and obtain a little iron in that way. So they have to be tied to posts on chains so short that their heads cannot reach the ground,

wrote Laurence Easterbrook in the *News Chronicle*, 4th June 1960. Veal farmers have tried to cure this habit, but of course cannot do so without giving more iron, and this, as we shall see, they cannot do, so the only solution is tethering.

Chewing straw would also develop the rumen which, as will be explained later, must not happen. It is claimed that it is easier to clean pens when the calves are on slats. The slats are brushed off once a day and the calf itself kept cleaner. The sloping concrete floor beneath the slats can be cleaned off into a tank and the manure easily disposed of. The farmer where there was straw, however, said that there was less risk of disease on straw as the calves were always warm, and that there was only complete cleaning out in between batches of calves and this was easier than cleaning out the slatted house.

I notice that one farmer has replaced his solid wooden slats by metal mesh to make cleaning still easier. The effect on calves has not, I think, been sufficiently tested to encourage many farmers to follow suit. The *Farmer and Stockbreeder* Vet. has stated that in his opinion slats are bad for stock, and I formed the impression that they were nervous and uncomfortable on them.

## Anaemia

It is now proposed to demonstrate that the veal calf as reared by present day methods is anaemic. This is a proposition that has been the subject of unrelenting and fierce controversy for years. In my opinion it goes to the very roots of the subject of veal rearing, for it raises the doubt as to whether the finished product is of any value as a food, and if it is not then the industry must of necessity wither away. The argument is unavoidably technical, complex, and I fear, somewhat lengthy, and in this you must bear with me. Even if there were some justification for rearing animals in the way described, and if there were some nutritional benefit to be gained, what justification can there conceivably be if there is no end benefit?

What is anaemia? The Oxford Dictionary defines it as lack of blood, unhealthy paleness. In medical parlance it is defined by the haemoglobin count of the blood, the lower the count the more anaemic the subject. Normally the iron-containing haemoglobin collects the oxygen and carries it to all parts of the body, which need it for metabolic purposes. We shall see that the veal calf is kept short of iron and therefore cannot oxidise properly. This lack of oxidation causes breathlessness and exhaustion which is disguised as long as it remains relatively immobile. In the extremity of anaemia the calf will cease to live, and it is by no means difficult to root out quotations which include the words 'drop down dead'. Death, apparently from anaemia and no other cause, has not been an unusual accompaniment to the past history of veal calf rearing and it would appear, on the evidence I have been able to collate, that the effort to produce the whitest of flesh involves rearing the calf in as anaemic condition as is possible whilst still avoiding this direst of results.

Anaemia is induced in the calf partly, no doubt, by its dim and sunless confinement, but mainly by the food it receives. The natural food for a new born calf is, of course, cow's milk. This has all the ingredients necessary for it *until it can begin to chew grass*, at the age of ten days to two weeks. Then from the grass and weeds in the fields it obtains the extra minerals needed to keep it healthy. It obtains vitamin D from the sun and grass. Even the ordinary calf reared indoors for the first three months of its life is given concentrates, water and ad lib. hay to supplement its milk feed.

The veal calf cannot, by nature of its value as a 'white' end product, be reared normally. We shall see that to produce white flesh the food is *designed to keep the calf anaemic*. The principal cause of anaemia is iron deficiency coupled with a deficiency of vitamin $B_{12}$.

The biochemist who carried out the analyses of the milk substitutes (referred to later) wrote: 'the B group vitamins ... are (normally) produced by microbiological action inside the animal. ... A definite deficiency of vitamin $B_{12}$ coupled with a low intake of iron will certainly result in an anaemia.'

The milk substitute goes directly into the abomasum or fourth compartment of the animal's stomach. 'When liquid ingesta contact the surface of the calf's throat, reflex closure of the lips of the groove occurs and the ingesta pass directly from the oesophagus to the abomasum, thus bypassing the other stomach compartment, namely the rumen, reticulum and omasum,' a veterinary surgeon explained in *U.F.A.W Courier*, Autumn 1960.

It is in the rumen that vitamin $B_{12}$ is synthesised. An essential aspect of the effort to induce anaemia is, therefore, to prevent the calf ruminating. In the normal calf 'as early as the first week or so, a limited degree of rumination and the urge to take up solids in small amounts occur,' the veterinary surgeon continued, and added that in veal calves 'the continued exclusion of solid and bulky foods from the diet results in the proportions of the stomach compartments being abnormal for the age of the calves, though *they retain their natural urge for ingesting foods in solid form.*' (The italics are mine.)

'Chewing the cud' seems an intrinsic part of the animal and it is difficult to visualise the calf without this natural and obviously pleasurable action.

A letter from one of the schools of agriculture explains the difference between a normal calf's diet and that of the veal calf:

> If the calves being reared to produce veal were fed on the same diet as those being reared for milkers:
>
> (*a*) They would not grow and fatten quickly enough to attain the re-required weight of 280–300 lb. live weight by the age of 12 weeks (after twelve weeks of age the flesh will darken whichever way the calves are fed);
> (*b*) The feeding of solid foods, e.g. hay and concentrates such as fed to the herd replacement calves would spoil the quality of the veal and cause the flesh to darken even before twelve weeks of age.

Mr Jennings, Past President of the British Veterinary Association, wrote:

> … There appears to be a belief that absence of iron in the diet is necessary to produce the whiteness, for we find many references to it in the literature. The following are some of the references:
>
> **1.** An article on Veal Production appeared in *Farm and Country* on 2nd September 1959. The article is by Dr Bakker who was then the Agricultural Attaché, Royal Netherlands Embassy and is now advisor to Messrs Christopher Hill Ltd., of Poole, Dorset. On page 80 Dr Bakker states, 'It is necessary to maintain a slight anaemia to obtain white flesh.'
> **2.** In a pamphlet issued by Messrs Christopher Hill Ltd., of Poole, Dorset, manufacturers of calf milk substitute, and called 'Denkavit feeding for veal', 23rd September 1959, it is stated that Denkavit is an 'iron deficient food'.
> **3.** The Dutch manufacturers of milk substitute for calves, Messrs Trouw and Co., have published a pamphlet called 'Fattening and Raising of Calves'. On page 8 it says, 'Avoid ferrous water.'
> **4.** Article in *Farmer and Stockbreeder*, 25th October 1960, page 64 – 'Veal Calves (Almost) in Comfort' by Mary Cherry. In this article it states 'Iron intake is strictly controlled. The iron content of the water is known and extraneous sources are eliminated; for example, all ironwork is galvanised. The aim is to give the calf its immediate iron requirement but to avoid excess which would be stored in the body and spoil the whiteness of the flesh.'

It is curious, in view of what I have just said, to find feeding firms, veal farmers and even officials of the Ministry of Agriculture, hastening to assure us that by feeding the calf up to three months on a milk substitute exclusively they are following the calf's 'natural diet'. The only time the veal calf gets a natural diet is when it gets colostrum from the cow during the first four days of its life and *before* being bought by the veal farmer. Even at that stage a calf is chosen which is potentially anaemic. Producers are advised by one of the feeding firms to choose their calves carefully:

> The calf should not have red gums or palate, nor should the corners of the eye show red. It is difficult to grow white fleshed calves when starting with 'red' calves. Also, under the tail should be checked for pinkness, rather than redness.

Is this following the old Dutch methods? The Agricultural Correspondent of *The Times*, 15th August 1960, told us:

> Veal production in the old days flourished mainly in districts where iron in soil and herbage was low and the dam passed little to the calf.

The milk substitute, I was informed by the manufacturer, is 'composed largely of milk powders together with added fat, vitamins and trace elements and the guiding principle behind its formulation is the composition of cow's milk'. The Ministry of Agriculture commented:

> The B vitamin content of properly manufactured milk substitutes is unlikely to differ very much from that of whole milk. The fat-soluble vitamins removed with the butter fats in the manufacturing process (vitamins A, D and E) are replaced by most manufacturers….

While Mr Jennings commented:

> Since the beginning of the controversy regarding the possible cruelty to calves there has been a tendency for a toning down of some of these statements (regarding anaemia) and many people including one or two well known personalities have been hoodwinked by some recent statements. It is now being stated by the manufacturers of calf milk substitutes that iron is actually added to these substitutes and that they contain more iron than does natural milk. Such statements are perfectly correct *but it is not the whole truth. It is well known that cow's milk has very low iron content and if young calves are fed on it continually they develop a very severe anaemia. It is, of course, not natural for a calf in nature to be fed entirely on milk for at a very early age the young animal begins to nibble at grass and thus obtains sufficient iron to avoid anaemia. At the time when the Dutch produced white veal by feeding cow's milk only, the calves would often drop down dead when they were released from the boxes in which they were confined and that, of course, was when they were taken out for slaughter*. (The italics are mine.)

Even a leading analytical chemist of animal foods has made the statement that there is a higher iron content in the milk substitutes than in normal dried whole milk and that it is therefore suitable for the purpose for which it is used. Of course it is suitable for the purpose for which it is used. But that is to make the calf anaemic, not to keep it healthy.

In Parliament we get the same facile remarks (*Hansard*, 25th July 1960):

> Lady Gammans asked the Minister of Agriculture, Food and Fisheries if he will introduce legislation requiring firms selling milk replacer in this country to comply with mineral nutritional requirements, bearing in mind the absence of iron in the water of some areas.
>
> Mr Hare: No: the proprietary milk substitutes used for veal production are largely based on dried skim milk and therefore are basically very similar in mineral content to natural whole milk. I understand that further minerals are added.

A paper read to the British Veterinary Association and reported in *The Veterinary Record*, 27th August 1960, contained the following paragraph, indicating that some veterinary surgeons also seem unaware of the inadequacy of the milk substitute as a complete diet for the calf:

> There has been some apprehension in the United Kingdom about the possibility of minerals being deliberately removed from the milk substitute that comprises the feed. Careful inquiry elicited the information that, in fact, for a variety of reasons calves fed on milk substitute are likely to receive more iron, for instance, than calves raised entirely on whole milk. Indeed, experience has shown that the general health of animals fattened on milk substitute is higher then than when whole milk is the diet. There are not the same disorders nor the same degree of anaemia and the loss of calves from sudden death is markedly less. There is evidence, however, that in areas where the iron content of the water which is added to the milk powder is low it is desirable to balance this with an iron additive for the well-being of the animal and the assurance of an economic return to the farmer.

You will notice that it is only claimed that anaemia is less, not that it is eliminated, and that the incidence of calves dropping dead is less, this also has not been completely eliminated.

Dr Bakker, speaking of the final production of Denkavit, a Dutch milk substitute, made the interesting remark: '… we have been able to *rear healthy calves, with a mild anaemia*, producing the high quality white veal the consumer wants.' (The italics are mine.) His firm, Denkavit, make a special point in their booklet describing their milk substitute designed for the first eight weeks of a baby beef calf's life that this is a completely different formula from that for veal calves:

> Denkavit Rearing is a full fat calf food processed and blended from premium ingredients, fully vitaminised and with a special mineralised complex for the requirements of replacement stock. IT IS NOT AN ADAPTED VEAL FOOD as are many others. Stock rearing and veal production have no similarity at all and therefore universal feeds are bound to fail.

Whilst assuring us that milk substitute is all that can be desired to keep a calf healthy, all the milk substitute manufacturers are curiously reticent in allowing us any information on its analytical content. This, we are told, is because their formula is secret, other manufacturers should not be allowed access to it, but it transpired that all formulas were known in the trade. Even the Ministry wrote, 'Details of the various products on the market are available from the firms concerned.' But the firms in fact resolutely refused their analyses. In the end we had our own analyses made of samples of milk substitutes by a leading Dutch firm and a leading English firm and it will be seen that there is no essential difference in them.

I have taken the amount of Dutch feed given at 60 days in the pen (into which it goes at about four days old), when it gets a little over 13 pints of milk substitute a day, that is, at a concentration of 1 lb. of milk powder to 6 pints of water, it receives 1 kilogramme of milk powder (2 lb. equals 0·907 kilogrammes). The English equivalent is for the 8th to 9th week when it receives the same.

At this time, at a starting weight of, say, 80–90 lb. at four days, and a conversion ratio of 1·4: 1, the calf will weigh 140–150 lb. These figures are, of course, only averages, as each calf varies in its starting weight and its ability to put on weight.

The following is an analysis of these feeds based on an intake at the stage we are considering of 1 kilogramme a day:

| Constituent | Dutch milk substitute | English milk substitute | Requirement of normal 150-lb. calf* |
|---|---|---|---|
| Moisture | 111 g. | 116 g. | |
| Protein | 175 g. | 250 g. | 280 g. |
| Fat | 164 g. | 118 g. | between 1 and 2 per cent |
| Carbohydrate and fibre | 550 g. | 457 g. | |
| Mineral matter | 60·5 g. | 59 g. | |
| Calcium | 10 g. | 7·7 g. | 11·0 g. |
| Magnesium | 1·1 g. | 1·3 g. | 3·0 g. |
| Sodium | 3·2 g. | 5·4 g. | 2·25 g |
| Potassium | 9·3 g. | 12·4 g. | |
| Phosphate | 17·8 g. | 3·4 g. | 9·0 g. phosphorus |
| Chloride (Cl) | 10 g. | 3·8 g. | less than 5 g. |
| Sulphur ($SO_4$) | 6·0 g. | traces | |
| Copper | 0·6 mg. | 27·0 mg. | 18 mg. |
| Nickel | 0·2 mg. | 9·0 mg. | |
| Iron | 34·0 mg. | 30·0 mg. | 225 mg. or 56 mg. per 100 lb. 84 mg. no storage |
| Cobalt | ·01 mg. | 2·0 mg. | 0·0 |
| Zinc | 8·0 mg. | 35·0 mg. | |
| Molybdenum | | 1·0 mg. | |
| Manganese | | 11·0 mg. | 37·0 mg. |
| Vitamin A | 200 I.U. | 1,300 I.U. | 5,000 I.U. |

*Continued*

Continued.

| Constituent | Dutch milk substitute | English milk substitute | Requirement of normal 150-lb. calf* |
|---|---|---|---|
| Vitamin B₁ | 0·2 mg. | 0·8 mg. | |
| Vitamin B₂ | 0·4 mg. | 3·0 mg. | 1·2–2·1 per 100 lb. |
| Nicotinic acid | 0·9 mg. | 9·0 mg. | |
| Vitamin E (total tocopherols) | 2·5 mg. | 5·0 mg. | 22·5 to 225 mg. depending on amount of unsaturated fat |
| Vitamin B₁₂ | 60 μg. | 20 μg. | 34–67 μg. 25 lb. per day water for 150-lb. calf |

*J.H.B. Roy, *Conference on the Scientific Principle of Feeding Farm Livestock, 1958*.

SUGGESTED DAILY NUTRIENT ALLOWANCES FOR GROWING AND FATTENING CATTLE, ROY 1958

| Body weight lb. | Dry matter lb. | Water lb. | Energy maintenance plus 2 lb. | Digestible protein lb. maintenance plus 2 lb. | Ca g. | P g. | Mg g. | Na g. |
|---|---|---|---|---|---|---|---|---|
| 100 | 1½ – 3 | 10–20 | 4,500 | 0·55 | 10 | 8 | 2 | 1·5 |
| 200 | 6 | 20 | 5,250 | 0·70 | 12 | 10 | 4 | |
| 300 | 8 | 30 | 6,250 | 0·85 | 13 | 12 | 6 | |
| 400 | 11 | 40 | 7,250 | 0·95 | 14 | 13 | 8 | |

Trace elements and vitamins, per 100 lb. bodyweight:
Cu 12 mg., Fe 150 mg., Mn 25 mg., Co 150 mg., Carotene 15 mg. (Vitamin A 3 mg.), Vitamin D 450 I.U.. Vitamin E 15–150 mg. (depending on amount of unsaturated fat in diet).

Whilst appreciating the fact that an immobile calf needs rather less nourishment than an active one, the analyses would still suggest from available data that the milk substitutes are low in protein, magnesium, iron, manganese, vitamin A and vitamin E. In addition, the Dutch milk substitute is low in copper and vitamin B₂, and the English in calcium, phosphate, chloride and vitamin B₁₂.

Dr Roy stresses that his chart is not a precise table of requirements but only a suggested table giving a margin of safety for the health of the animal.

On the inclusion of minerals in the food he comments:

> In assessing the allowance for minerals, considerable margins of safety may be required over minimum requirements owing to the interactions that occur among minerals in the digestive tract, whereby one mineral may precipitate another in insoluble form. For instance, excess of calcium may impair the assimilation of both iron and iodine. … The trace minerals known to be essential for cattle include copper, iron, manganese, cobalt, zinc and iodine.

J. O. L. King says in his book *Veterinary Dietetics*:

> If calves are raised mainly on milk, without much roughage, anaemia may develop in rapidly growing individuals because of a deficiency of iron and copper. Anaemia can be prevented by providing small amounts of these elements as specified (by Roy).

The trace mineral which is startlingly low in the milk substitutes is iron. In assessing how much iron a young calf needs daily Roy writes:

> Iron: whole milk contains only about 2 mg. Fe per gallon and calves restricted to such a diet develop anaemia. Blaxter, Sharman and MacDonald (1957) have estimated that to maintain normal haemoglobin (Hb) values and to achieve adequate liver storage the net requirement of Fe is about 25 to 50 mg. a day for weight gains of 1 and 2 lb. per day, respectively. Matrone et al. (1957) calculated that a 500-lb. calf needed 1·2 mg. Fe a day to maintain a blood value of 10 g. Hb per 100 ml. and a further 16 mg. a day for a weight gain of 2 lb. per day. Their gross requirement, which contained no allowance for normal storage of iron, would, assuming 30 per cent utilisation (Matrone et al., 1957), be about 56 mg. per day for a calf gaining weight at 2 lb. per day, compared with a value of about 166 mg. per day based on the calculations of Blaxter et al. (1957). It would seem, therefore, that an intake of 150 mg. Fe per day should fully cover the requirements of calves gaining weight at the maximum rate. Furthermore, Thomas, Okamoto, Jacobson and Moore (1954) found that 100 mg. Fe per day increased the red blood cell and haemoglobin values of calves that were moderately anaemic.

Mr J. Wilson, who was at one time with F.M.S. (Farm Supplies) Ltd., wrote an article on veal rearing for *The Veterinary Record*, 2nd December 1961. He felt it incumbent on him to mention the controversy then raging in the press about anaemia in veal calves and explained his firm's position:

> Whole milk contains only about 2 mg. iron per gallon, but few veal calves reared on whole milk die from iron deficiency anaemia. The reserve of iron in the body at birth must, therefore, be quite high, but it will be influenced by the diet of the dam. During the drying process the milk by-products take up a certain amount of iron and the milk substitute with which the writer is personally familiar contains 30 mg. iron per kg. dry matter. The limited information available on the iron requirement of the calf suggests a figure of 150 mg. a day for a weight gain of 2 lb. per day (Blaxter, Sharman and MacDonald, 1957).
>
> This figure assumes a 30 per cent utilisation and also provides for the maintenance of normal haemoglobin values and adequate liver storage. The gross requirement with no allowance for normal storage is estimated to be about 56 mg. per day (Matrone, Conley, Wise and Waugh, 1957). Even if a gross iron deficiency anaemia were desirable – which it is not – it is difficult to see how such a condition could be produced in 12 weeks when feeding a milk substitute containing 30 mg. Fe per kg. dry matter. Carcases from calves reared in metal pens did not differ in any way from those of calves reared in wood pens and fed a similar ration.

We have seen that the veal calf, because of its early slaughter, needs no storage of iron in its liver. It has usually, depending on the diet of its dam of course, sufficient iron in its liver when it is born to last it up to six weeks of age. After that age it is completely reliant on what it receives. Loss of appetite and listlessness are often reported in veal calves at about this age and these are usually associated with iron deficiency. This is the age presumably when anaemia first sets in.

We have seen that with no storage it should have 56 mg. a day per 100 lb. bodyweight. At 150 lb. bodyweight, therefore, it should have 84 mg. a day. Our two analyses show a daily intake of 34 mg. and 30 mg., and Mr Wilson's firm gives a figure of 30 mg. a day, instead of the recommended amount of 84 mg. How can his argument be justified?

Experiments were undertaken for the Ministry by the National Institute for Research in Dairying and when taxed as to the results of these experiments the Ministry at first denied anaemia or the need for blood tests:

> No blood examinations were made of the calves used in the Ministry's trials, nor was there any reason to suspect from close and constant observations of these animals, their appearance and, above all, their live-weight gains that they were in any way anaemic….

That was in 1960, but in 1961 they wrote to an M.P.:

> Experiments in veal production have been carried out by the National Institute for Research in Dairying in collaboration with the Royal Veterinary College. Blood tests of calves fed entirely on whole milk and various milk substitutes throughout the fattening period have indicated mild anaemia.

It was not, however, until 20th December 1962 that they proferred the information to a friend of mine:

> The blood tests taken on the veal calves were made at regular intervals between birth and slaughter at approximately 250 lb. live weight. There are no universally accepted degrees of anaemia other than its measurement in terms of blood haemoglobin. In the experiments previously referred to, carried out by the National Institute for Research in Dairying in collaboration with the Royal Veterinary Colleges, the minimum mean haemoglobin values at slaughter varied between 7·0 g./100 ml. and 5·0 g./100 ml.
>
> These results need, of course, to be assessed in the light of such circumstances as the amount of exertion required of an animal.

and further explained on 15th March 1963 the type of anaemia:

> The level of haemoglobin in calves on high-level liquid feeding regimes normally falls from its initial level of about 12 g. per 100 ml.: the level at slaughter varies between 5 and 7 g. per 100 ml., according, to diet. The anaemia produced by the diet is a simple nutritional anaemia which responds to iron supplementation.

Dr Reginald Milton, B.Sc., Ph.D., F.R.I.C., M.I.Biol., the biochemist who carried out the two analyses, wrote after completing the 'Dutch' analysis:

> Although the food is rich in bone-forming minerals and is satisfactory with regard to content of most trace elements, it is certainly deficient in iron. It is also deficient in all vitamins of the B group and Vitamin A. Calves normally carry reserves of the B group vitamins present at birth until they can augment milk with grass, hay, etc. Then the stomach activities become modified and these vitamins are produced by microbiological action inside the animal. If such material is excluded from the diet (as under the broiler method) then Vitamin B deficiencies of all kinds will eventually ensue. A definite deficiency of Vitamin $B_{12}$ coupled with a low intake of iron will certainly result in an anaemia….
>
> I am of the opinion therefore that this product… fed under the conditions recommended would produce in the calf chronic anaemia and also other vitamin deficiency symptoms.

A distinguished veterinary surgeon has stated that in Holland the veal calves would sometimes drop dead from anaemia when they were taken out of their pens for slaughter:

> The calf food manufacturers therefore add some iron to the substitutes but only in sufficient quantities to prevent such severe anaemia that the calves drop down dead,

and he went on to say that an animal nutritionist attached to a veal food firm had stated at a meeting of veterinary surgeons, "We must control this anaemia." This statement means not only must the excessive anaemia be controlled but that an excessive iron intake must also be controlled. It is well known that iron can be given in excess in the diet, and anything not required by the body is excreted in the intestine. (This official from the veal food firm) spent a lot of time testing the water in all parts of southern England. Such testing is obviously not necessary to find the minimum amount of iron because it would be much easier and safer simply to put excess iron in the food. The well testing is to see that the water does not contain too much iron.'

A point arises which I find interesting. In the Ministry of Agriculture trials on veal calves in seven of their experimental farms to see whether there could be obvious differentiation of carcases of calves fed on whole milk or on milk substitutes, for purposes of deciding on the possibility of a subsidy for veal, none could be found (*Experimental Husbandry No.* 5, Ministry of Agriculture). We have been assured so many times that the calves fed on whole milk were so acutely anaemic that they often died; does it follow that if the milk substitute carcases are of the same colour they are of the same degree of anaemia? Dr Shillam of the National Institute for Research in Dairying, wrote in *Agriculture*, September 1961:

> The milk substitutes used in Holland are low in iron; this is because the water, often obtained from wells, is rich in this element.

As, however, the water supply in the southern part of England at least appears to contain little or no iron, some should be added to the milk substitutes used in this country as a safeguard against the development of clinical symptoms of anaemia, particularly where calves are housed in wooden crates.

But, he had pointed out,

> The largest single factor at present determining the price paid to the producer appears to be the so-called whiteness of the flesh, and here little can be done except perhaps to ensure that the calf is not allowed access to unlimited supplies of iron from parts such as rusty bucket rings and gates, or from solid foods such as concentrates.

And I quote him finally as saying:

> The high energy milk substitutes are nicely balanced in their content of iron so that on the one hand there is sufficient to prevent the calves showing clinical symptoms of anaemia, and, on the other, there is not too much to make the carcase unduly red. (*Farmer and Stockbreeder*, 24th October 1961)

Perhaps I should, to close this case, refer you back to the statements on Dutch veal calves which were said to look perfectly normal until the time they suddenly dropped down dead.

They were clinically anaemic!

In Holland the use of hormones as a growth stimulant is allowed for veal calves, and antibiotics are also freely in use for preventing disease thereby allowing uninhibited growth.

> Antibiotics may be included in milk substitutes used in Holland, and there is a large body of opinion which considers that antibiotics are essential for successful veal production. Their use in this country would probably make veal production more likely to succeed, especially under unfavourable husbandry conditions, provided that there is no danger from strains of organisms developing resistance to antibiotics. (*Agriculture*, September 1961)

The use of antibiotics for veal calves has at last been permitted in this country. A proposal to left the ban was put forward in the autumn of 1962. Even before this proposal, however, veal farmers were allowed to use antibiotics supplementary to the calf's feed to suppress disease. Now presumably they will be able to use them more freely.

'Whoever told you that veal calf rearing entails no losses,' a veal farmer exclaimed to me, 'was lying. We have great trouble in keeping them alive.'

We have seen that veal calf rearing is a continual process of trial and error. Misconceptions about degree of lighting still persist, and opinions on this vary from holding to holding. Dr Bakker has told us that the idea that darkness was necessary probably originated from the fact that plants grown in the dark are pale, prisoners are pale and so on. Complete darkness is not now considered necessary although until now it was considered vital in Holland. Dimness is, however, advocated for various reasons, the chief one being that calves should lie quietly and might waste energy in being restless in bright light or sunshine.

> There is no evidence whatsoever to show that darkness favours the production of white flesh, and provided direct sunlight is avoided the animals will keep just as quiet and rest in normal light. (*Agriculture*, September 1961)

Another reason given is that darkness discourages flies, but at the Dutch Institute for Animal Husbandry and Meat Production it was found that calves housed in full light had a slightly better food conversion ratio than those kept in darkness. Again there were more than twice as many flies in the dark sheds as in the light.

A caption to a photograph in the *Farmer and Stockbreeder*, 13th September 1960, reads:

> Flies are everywhere. Look closely at this picture and you will see them lying dead in the gutter in their hundreds. Resistance to all known fly sprays is said to be developing in Holland.

I will have more to say on this subject in another chapter.

A veterinary surgeon has confirmed my impression of the misery of calves kept in solid-sided crates in the dark by telling us that 'it is a fact, however, that if left in total darkness, calves fret and become apprehensive and lack assurance that they are part of a herd.' (*U.F.A.W. Courier*, Autumn 1960.)

Dr Bakker, writing in *Veterinary Record*, 27th August 1960, on veal production, said:

> Calves are herd animals and do not feel safe and happy unless they are with other calves; this appears to be very important. If you enter one of these modern stables where calves can see everything they will not take very much notice of you; even if you kick over a bucket there is no question of sudden fright.

While a slaughterer told me that the only time calves became terrified in the slaughterhouse lairage was when they were among the last to go to the killing. While they were in a large group they did not seem to notice the constant removal as calves were taken.

Perhaps, with some farmers, deficiency in diet is another misconception, they may not realise the full intricacies of the reasons for this, or indeed that they are deficient.

> Many farmers are unaware of the fact that some veal foods are lacking in certain minerals and vitamins (I was told by a veal farmer). They use a manufactured food which is sold to them in many cases by high pressure salesmen and advertising, just in the same way as we are sold food; we do not ask for a detailed analysis.

After this barren and boring half-life the veal calf is taken to the slaughterhouse. It cannot, like other animals, be left in the lairage for longer than eight hours because the slaughterman would not be able to meet its specialised feeding requirements, so it is slaughtered as soon as possible after arrival. In Gentile slaughter it is stunned either by electric stunning or by the 'humane stunner', decapitated and then inflated with air subcutaneously to give the small amount of fat an 'attractive' appearance. In Kosher slaughter it is hoisted by a back leg, has its throat cut and hangs until it bleeds to death.

And what of the finished product?

Is it really worth it?

According to their Agricultural Attaché, in Denmark the public don't think so. They prefer to eat the rather redder veal from naturally reared calves.

It is interesting to note that even in Holland itself with

> less emphasis on whiteness of flesh as a criterion of quality an extreme degree of restriction is hardly necessary. The Dutch domestic market, it seems, is less insistent on white meat than some of the countries to which Dutch veal is exported. (*The Times*, 15th August 1960)

A butcher told me that he did not think that one butcher in a hundred ate Dutch veal, they preferred meat of quality, and a farmer friend told me that she suddenly decided to have a four-month-old calf she was rearing slaughtered, and the 'veal' they ate from it was the tastiest and tenderest she had ever eaten, even if the flesh was a shade darker than that of calves reared for veal. This was what is known in the trade as a 'runner' and bears out the findings of the following experiment in Wisconsin:

> The largest single factor at present determining the price paid to the producer appears to be the so-called whiteness of the flesh (wrote Dr Shillam in *Agriculture*, September 1961).... The desire for 'white' flesh is not, apparently, based on cooking quality, taste or tenderness. In fact, experiments in Wisconsin have shown that supplements of iron and copper given to milk-fed calves produce a more tender veal. Whiteness seems to be of value only in that it gives the meat an attractive appearance and shows the consumer that the joint is from a milk-fed veal calf, and not from an indifferently reared animal.
>
> The ration has this white flesh shibboleth as one of its principal purposes (*Farmer and Stockbreeder*, 13th September 1960), 'The housewife must have white veal,' they say, 'she will not buy red veal.' So be it.

I will let R. Trow-Smith, writing in *Farmer and Stockbreeder*, 13th September 1960, of these Dutch methods which he saw on a visit to Holland, have the last word: 'I would not keep a calf this way, I like to get pleasure as well as profit from my stock'.

# Other Intensive Units

I have described the lives of broiler chickens and battery hens in some detail because they represent two large and established industries, but a vast number of animals are kept on these factory farms and the list is continually being extended. Turkeys, ducks, quail, rabbits, pigs and beef calves are also included, and experiments are on foot to include even lambs. It seems that we might lose even this joyous symbol of spring from the countryside.

The veal industry has deserved detailed discussion because the precision with which the calf has been deprived of both comfort and health is almost unbelievable, and although in comparison to the chicken industries it might seem small, some of the niceties of the system are being extended to beef calves.

## Broiler Beef

It appeared at one time that beef calves were safe from the push to intensive production by virtue of their reputation for producing a 'quality' meat which was not open to the standardisation at a lower level such as had befallen the chicken. But we reckoned without the supermarkets who claim that the housewife is forcing them to buy a 'cheap, uniformly pale-pink sort of beef that can compete with the broiler chickens and frozen fish in their food display cabinets; the stuff has to start pale pink, as meat darkens when exposed' (*The Observer*, 19th August 1962). Whereas it used to take from two to three years to rear a beef calf, grazing on the

© J. Harrison and J. Wilson 2013. *Animal Machines* (Ruth Harrison)

best available pastures, it takes only eleven to thirteen months to produce the type of calf to fulfil this supermarket demand. These are lighter animals, killed at around eight and a half hundred-weight, but to achieve this weight in the time some degree of restriction is necessary.

Systems vary from farm to farm. Many farmers still believe in allowing the calves to graze for at least part of the time, others confine them to yards or large airy barns open at one or both ends. But the call of broiler profits becomes increasingly strong and the tendency is to put the calves into controlled environment buildings and to restrict movement by heavy stocking. The Rowett Research Institute, who have pioneered 'barley' and broiler beef in this country, based their conceptions on the American feed lot. John Cherrington, writing in the *Financial Times*, 21st June 1963, thought they had done this 'rather erroneously.... In America the cattle fattened are pure bred beef cattle ... bred on the open range, and only put into yards for the last four months of their lives.... It's important to remember that store cattle in America have grown their basic frames under cheap range conditions, and have been weaned on their own mother's milk.' John Cherrington contrasts this with our system: 'the calves are never allowed out, and in many cases don't even lie on straw in case in their desperate search for fibre, they eat straw and so dilute their concentrated feeding. Mostly they are kept on wooden slats.' Our feed lots are designed to hold up to two thousand calves at a time. Already one firm has achieved this number, whilst feed lots of five hundred or a thousand calves are becoming a regular occurrence.

Where slats are used it is reckoned that double the normal rate of stocking can be employed, and some units have taken to tethering the calves at 2 ft. 6 in. centres. There is a great danger that the frenzied 'food into flesh' techniques which have enveloped the veal calf will all spread to these unfortunate beef calves. The comments in the *Daily Mail* by Laurence Easterbrook, after he had visited the Royal Show in 1963 during which he discussed barley beef with producers, suggest that some of the methods ape exactly the veal conditions described in the last chapter:

> The calves are put in little cells at three days old and never leave them until they are 11 months and ready for the butcher.
>
> They have no straw. They stand on slats through which their manure drops to a pit.
>
> Even at maturity they must live their lives in a space of 17 sq. ft., say 8 ft. × 2 ft. 3 in.
>
> I talked to a man who had seen this place in operation. 'It is diabolical,' he said, 'they hate the slats. They slip on them and fall on their knees. The stench after only one month is indescribable....'

A *Farming Express Supplement*, December 1961, described the advantages to the farmer of a push-button system for beef production:

… Feed systems are of different types, but are based mainly on an auger running the length of the feed troughs.

As many as 300 bullocks can be fed in 20 to 30 minutes using such systems. All the stockman has to do is to start and stop the machinery.

A logical extension of the automatic feed system is to put the animals on slats or use a scraper to take the dung from the front of the feed troughs to a tank. Here it is mixed with water to form a slurry. The liquid is put back on the fields by means of the rain gun irrigation system so completing the cycle.

*The Financial Times*, 9th May 1963, reported that a special prize was awarded in a competition organised by the Country Land-owners Association for new beef houses, to a building which the judges said was 'perhaps the prototype of things to come; the most challenging of all entries, *incorporating extreme intensity and complete detachment from organic farming*'. (The italics are mine.)

The building is modelled on the lines of poultry broiler houses. Calves are installed at three days old and stay in until they are ready for slaughter at 11 months. The building needs no litter, merely a slatted concrete floor with mechanical feeding and waste disposal.

Dr Preston, of the Rowett Research Institute, envisaged automatic feeding being operated by a time switch thus eliminating the need for weekend work.

An ideal system for the farmer, and the stockman who does not care about stock as stock, but regards them merely as expendable profit-producing machines.

Feeding firms sell not only food for livestock these days, but systems of rearing round their feed. A hazard of this is that they lure too many customers into a particular branch of farming and cause profits to fall, but with calves this is a slight hazard as cows cannot be made to produce calves at the speed with which chickens are produced.

Most feeds for feed-lot calves are based mainly on concentrates with little or no access to roughage. Where slats are not used, sawdust is advised rather than straw, which the calves might chew.

The Rowett system, evolved by Dr Preston at the Rowett Research Institute, and now very widely used, gives the calves a diet almost exclusively of barley, with added minerals and vitamins, antibiotics, tranquillisers and hormones. It is considered that hay or other roughage would slow down the rapid weight-gains made by these 'barley-fed' beef calves. Hormones are estimated to increase the calf's weight by 15 or even 25 per cent, and are fed with the concentrate for the last three months of the calf's life. I will be discussing this further in a later part of the book.

Tranquillisers are given to the calves as a routine measure to keep them quiet, and some beef calves never see the light of day but are kept in 'darkened pens to boost their growth and keep them relaxed' (*Daily Express*, 6th September 1962).

*Farmer and Stockbreeder*, 4th April 1961, reported Dr Preston as saying that 'niceties such as linseed cake were never given in the Rowett trials; a shining coat did not bring in any extra profit'. Recent reports are beginning to show that not only do these calves lack shining coats, but some are coming out blind, many have damaged livers, and all suffer from a degree of pneumonia. Yet all are top graded at the slaughterhouse.

Probably the chief factor holding producers back from broiler beef at the moment is that the price of barley might go up and the price of calves might also go up. Friesians have been found to put on weight in these conditions more quickly than other breeds, and there is likely to be a shortage of Friesians. Let us hope that these hazards will last long enough for farmers to be enabled to make a deeper study of the effects of these methods on the animals before plunging even deeper into this form of intensivism.

## Rabbits

The broiler rabbit industry began in 1959 and has quietly crept up to a production rate in 1963 of fifty-two million rabbits a year. Broiler rabbit meat is sold by butchers and supermarkets, and its characteristic paleness and tenderness allow it to share the 'quality' hallmark of broiler chickens, veal and beef.

The industry had first to find does with characteristics yielding good potential profits. The most important of these was, of course, 'liveability'. It was no use taking a doe, however large a litter she might have, if she did not survive long enough to breed many times. Then strains had to be developed which were good at fulfilling the main task of intensivism today, converting food into flesh with a good killing-out percentage. A rabbit has also to have a good pelt. The fur from these broiler rabbits is mostly used by the hat trade and a coloured pelt fetches only fourpence while a pure white one will fetch a shilling.

Although rabbits had not previously been reared intensively for meat, they had been reared under similar conditions for a great many years to meet the demands of experimental laboratories, and thus the industry had much useful information on which to base its research. Nevertheless, mortality in the cages is very high and it is reckoned that one in three of the growers dies before reaching the required weight of about 4 lb. in eight weeks.

A rabbit unit on first sight looks rather like a battery hen unit. Cages, each 3 ft. × 3 ft. or 4 ft. × 2 ft., with solid sides and back, hold a doe and her litter. Pellets are held in a trough in front of the cage and water is sucked from a drip valve. The removable nesting box has a solid floor with some straw for the comfort of the doe and the tiny newborn rabbits which are born hairless, blind and deaf. When the nesting box is removed, all exist on wire mesh which, although provenly less comfortable for the rabbits, has the advantage for the producer that droppings fall through it and thereby keep the cages cleaner and save labour. On the other hand:

> General activity is apparently less on a wire floor and this may possibly be reflected in the health of the adult. Sore hocks are more common. It is suggested the so-called 'mucoid enteritis' may also be a greater hazard. Conception is said to be lower in winter than on solid floors as does are less receptive.
>
> Solid floor hutches are more comfortable, there is more general movement and sore hocks are not generally a problem. Greater attention to hygiene is required with fairly frequent additions of bedding. The continuous administration of a coccidiostat is probably essential. The winter conception rate is higher than on wire floors.

This article in *The Veterinary Record*, 25th November 1961, then suggested a compromise:

> Rabbits may be reared on wire floors and transferred to solid floors for breeding.

This is in fact what some units do. The litters, averaging eight, are early weaned at between three and four weeks and transferred to colony pens with other litters, where about sixteen growers are reared per pen, while the doe is given service again two or three days after the litter has been taken from her. She has to fulfil the abnormally high breeding programme of around ten litters in just over two years. If she does not do this she is not considered to be profitable. It is interesting to notice that a buck also has to earn his keep:

> To some extent, the amount of food given to the buck is dependent on the number of does they cover. Hay and water are available all the time, but the rest of the ration may be restricted or given ad lib. according to the breeding performance. (*Poultry World*, 1st March 1962)

Darkness is not recommended for rabbits. If there is no daylight in the unit artificial light is provided.

Denmark, where feelings ran so high that battery cages for hens were abolished, has a flourishing broiler rabbit industry and has had for a hundred years.

## Pigs

The pig is perhaps the most maligned of all farm livestock. In reality it is a scrupulously clean, lively and intelligent animal.

What is not generally understood is that the pig has a tough skin with inadequate sweat glands. Furthermore the thick layer of fat enveloping it acts as an insulator and makes it difficult for it to maintain its low body temperature against a high atmosphere temperature. Where inadequate facilities are available the pig therefore finds water to moisten its skin or wet mud to protect it from the sun. A dirty pig therefore is a reflection, not on the animal's innate characteristics, but on the quality of stockmanship and understanding of the farmer.

The Danes were the first to recognise that the pig must have a floor to rest on and a separate dunging passage, and the Danish piggery achieved an international reputation.

The present tendency to keeping pigs under artificial, intensive conditions, even though it is not a general indictment of modern pig keeping, points all too clearly to the fate that could await the domestic pig, which is considered by some farmers only in terms of the efficiency with which it can convert food into flesh, and the disappearance of affection for it has been accompanied with the disappearance of some of the niceties of its housing on the grounds of unprofitability.

The first thing to go was a separate feed trough. It was discovered that there were many advantages in scattering feed on the floor of the pen. There was no bullying as there was plenty of room for all to feed alike, the floor was kept scrupulously clean by the pigs and saved labour for the producer, and, perhaps the most telling factor, far more pigs could be got into the same amount of space because trough space per pig was no longer the governing factor. An experiment at Nottingham Farm Institute, for example, showed that in a pen 11 ft. × 6 ft. only eleven pigs could be housed with a feeding trough, but when floor feeding was tried it was established that thirty young pigs, or between eighteen and twenty adult pigs, could be housed.

> This high level of stocking has been successful although a careful watch has to be kept on the pigs to ensure that none is forced to sleep outside in the dunging yard due to overcrowding,

stated the report in *Farmer and Stockbreeder*, 22nd January 1963, and later added:

> … A few pigs have died from unexplained reasons which might be due to the stress conditions associated with high density stocking. These deaths in no way nullify the extra return obtained from the higher total output.

It seems that the atmosphere in the pig houses is so dusty that it is uncomfortable for the pigman, and it is emphasised that only virus pneumonia free pigs can survive under the conditions. Again and again is stressed the importance of outdoor rearing until weaning time, to give the piglets the hardiness to withstand their future conditions.

Once more science has come to the aid of the pigman as a partial replacement for good stockmanship. It has been found that by hysterectomy, or the removal from the sow of the whole uterus with the piglets inside it, and their removal and maintenance in sterile incubators for fourteen days, they can be sent onto farms at the age of five weeks and will be free of infection as long as they are not in contact with other pigs. These pigs can then make the basis of a new disease-free herd. Hysterectomy originated in America where the standard of stockmanship with pigs was notoriously low. In this country stockmanship has been much better and disease could be controlled by improving it still further. But is this being done?

Besides eliminating trough feeding, the elimination of the dunging passage has been the next big space-saving move. The needs of the pig have been met by either a dung channel down the side of the house covered by slats, or by making the floor slightly sloping and encouraging the pigs to dung on the lowest part of the slope from where the dung runs through pipes into a sludge tank on the outside of the building. This last is the method used most widely in Northern Ireland, where they call them the 'sweat-box' piggeries. Sweat-box piggeries have been fairly widely tried in this country. Insulation of the buildings has been perfected to such a degree that the intense heat generated by the pigs themselves keeps the temperature in these units at around 80°F. or even higher. There is no forced ventilation, the only ventilation being through the top half of a stable-like door. At this temperature it is estimated that conversion of food into flesh is most economical. And indeed the pigs have little inclination to do other than eat and rest. The sweat-box pigman has found that this no-trough, no-dunging passage, no-bedding, no-cleaning out routine has brought him big profits and moreover has meant that one man can look after upwards of a thousand pigs. The pigman's chief glory, however, is that the pigs remain healthy – or rather disease-free. This is easily explained. The atmosphere in the unit is humid, for the damp from the urine and the pigs' sweat rises up in the intense heat taking with it all bacteria, which find their way out of the houses with the steam, or remain caught in stalactites hanging from the ceiling.

This consequently becomes the easiest way to rear unhealthy pigs. And if some farmers feel squeamish about keeping their pigs in this way, let them hear the opinion of Dr K. C. Sellers, director of the Animal Health Trust's farm livestock research centre, who was reported by *Farmer's Weekly*, 23rd February 1962, as having

> … pointed out that pigs were kept to make money, as carcases, and one should not get over-sentimental about them.
>
> 'I think the test of a "sweat-box" is whether or not it pays. If your pigs are healthy, then it is not a worthwhile proposition.'

In fact research has proved that pigs grow more slowly this way as they eat less, and some farmers have proved that it is still more economic to rear them outdoors. More economic and a happier way of doing it.

Whether in 'sweat box' or more natural pig units, advice to producers these days remains to 'cover the floor with pigs' and some advice given allows even less than the five square feet per pig until recently considered the maximum intensity. $3\frac{1}{4}$ sq. ft. per pig has been mentioned as being more profitable.

Heavy stocking leads to boredom and 'vice', in this case tail-biting, and producers are advised to throw a block of wood into the pen, or hang a chain in the middle of the pen to relieve the monotony. Another inevitable development has been deprivation of light. Darkness means no fighting, just resting and getting on with the job in hand, conversion of food into flesh. A description of a modern unit in *Farmer and Stockbreeder*, 26th March 1963, stated:

> The pigs are kept in semi-darkness. A 15-watt red bulb gives enough light for the pigs to see where to eat but not enough to allow fighting. Temperature and ventilation control coupled with the semi-darkness ensures that the meal is not wasted in unnecessary energy….

I would like to bring a gust of fresh air into this discussion of intensivism by quoting a letter written to *Farmer's Weekly*, 7th November 1961, by the director of a feeding firm:

> In the last war, I hired a derelict house and farm building and put about 100 pigs therein. Part of one wall in the house had collapsed but the staircase was intact and upstairs there was the bedroom to which the pigs had access. The pigman reported that there seemed to be competition for the bedroom every night and that in the daylight hours they would chase each other up and down the stairs.
>
> *I never had pigs do better than that lot.*
>
> I have come to the conclusion that our stock need variety of surroundings and that gadgets of different make, shape and size should be provided and that, like human beings, they dislike monotony and boredom.

> To sit in solemn silence
> In a dull, dark dock,
> In a pestilential prison,
> With a life-long lock,
> Awaiting the sensation
> Of a short, sharp shock,
> From a cheap and chippy chopper
> On a big black block!
>
> With apologies to W. S. Gilbert

# The New Factory Farming — A Pictorial Summary

The new production line methods of rearing farm animals are of deep concern to each one of us. Farmer, slaughterman, middleman, shopkeeper, housewife, all contribute to the chain, although some contribute in ignorance. Three important questions arise for us to consider.

How far have we the right to take our domination of the animal world – *in degrading these animals are we not in fact degrading ourselves?*

We have condemned them to an existence in which almost every instinct is frustrated and natural pleasure nearly eliminated; we do not allow them to live before they die. *At what point do we acknowledge cruelty?*

These animals are themselves unhealthy, and the drugs used so lavishly to keep them alive and make them put on weight speedily can have repercussions on man himself. *Can these unhealthy animals possibly make healthy human food?*

It must be emphasised that nearly all these photographs were taken by flashlight since the animals eke out their existence in darkness. The real gloom of their conditions can but rarely be captured in a photograph.

 © J. Harrison and J. Wilson 2013. *Animal Machines* (Ruth Harrison)

**Figure 1.** This veal calf dragged out its existence in a crate barely large enough to hold it and in the dark. It saw light only twice a day at feeding time. When the shutter was first let down its face was a picture of misery. We did not ever see the calf immured in the crate on the right.

**Figure 2.** Whatever its supposed inefficiencies, the traditional farm has contributed to the visual pleasure of the countryside, and one cannot help feeling that it is also a pleasant environment for the animals. On the good traditional farm there is a sense of unity between the farmer and his stock, he is a farmer because farming is in his blood, and profits are a secondary, if important, consideration. He recognises the animal's rights as a living creature. He recognises also that it is only healthy animals which can produce healthy food. He has worked hard to provide this and has earned the warm regard embracing the term 'the farmer's image'. *Photograph: Ronald Goodearl*

**Figure 3.** The new type of farm is like a straggling factory. The buildings jar on the eye and rob the countryside of much of its charm. These long sheds are completely utilitarian, each with its giant feed hopper to meet the needs of the animals permanently enclosed within. The new type of farm is a factory run on completely commercial lines by people who are business men rather than farmers. Their production line methods of rearing animals have put evolution into reverse and depressed the animal back towards the plant stage in an effort to turn it into an efficient food-into-flesh converting machine. With the increasing disappearance of animals from the countryside our children lose a very precious heritage.
*Photograph: by courtesy of C.A.A.C.A.*

**Figure 4.** This is a 'controlled environment house'. You will see them sprouting up all over the countryside. In detail they may vary slightly but in principle they are all the same. Controlled environment means complete isolation from the outside world. Vents down the sides of the house take the place of windows and the inside is lit only by artificial lighting. This means that specific lighting patterns may be followed, or the animal kept in virtual darkness as is so often deemed necessary these days to prevent the 'vices' which immobility, overcrowding and boredom can lead to. The temperature is thermostatically controlled, but it is worth noting that in few cases are the ventilation fans capable of keeping the temperature down during a heatwave and refrigeration is too costly for most of these units. Actual housing conditions for the animals vary. Broiler chickens and some laying hens are kept on a layer of shavings, called deep litter. This is turned frequently to keep it dry but only changed in between batches of birds when they are sent for slaughter. Other laying hens are kept on wire mesh floors but are also free to move around the house. A large proportion of laying hens and rabbits are kept in tiers of battery cages, their feet having only wire mesh on which to rest. Pigs are kept on an insulated concrete floor with no other covering and calves are kept on slats which may be of wood, concrete or metal. These methods are all aimed at saving the farmer the need to clean out, which is done only as each batch of animals goes for slaughter, but it does mean that in many cases the animal is condemned to live continuously over its own dung. More complete isolation from the possibility of any outside disturbance is sometimes achieved by relaying continuous light music in the shed. *Photograph: by courtesy of C.A.A.C.A.*

**Figure 5.** The interior below is of a broiler rabbit unit. Rabbit cages are usually 3 ft. × 3 ft., or 4 ft. × 2 ft. with solid sides and back. There is a nesting box with straw to one side of the cage in which the doe has her young. The white trough in front of each cage holds pellets and water can be sucked from a drip valve. Each doe has an average of eight in a litter but a third of these die before reaching the required age for slaughter of eight weeks, at which age they average about 4 lb. Pelts are used mostly by the hat trade. *Photograph by courtesy of 'Farmer and Stockbreeder'*

**Figure 6.** This photograph, and that a few pages further on, show the interior of typical broiler houses, both taken when the chickens were between five and six weeks old. It is not difficult to imagine the overcrowding at nine weeks when they are that much bigger and ready for slaughter. Each chicken is allowed a maximum of 0·8 sq. ft. on average, the size of a sheet of foolscap paper. The atmosphere in the broiler house has been described in a Ministry of Agriculture pamphlet as being 'dusty, humid and charged with ammonia' and this strikes one forcibly on entering the building. The chickens live from the age of six weeks in virtual darkness to prevent the 'vices', inevitable in such conditions, of featherpecking and cannibalism. It also ensures that no energy is wasted on movement. Note the hanging food hoppers, often automatically filled, and the water pipes mostly masked by the massing of the chickens. Other ways of preventing the chickens harming each other (see opposite) are by de-beaking, or by fitting opaque 'specs' which prevent the chicken from seeing directly in front of it. The first, and last, time the chickens see the light of day is when they are bundled into crates to be taken to slaughter. *Photograph: Dex Harrison.*

**Figure 7.** *Photographs: (middle) by courtesy of 'Associated Press'; (below) by courtesy of A.C. Moore.*

**Figure 8.** Crates are piled up one wall of the slaughter room in the packing station in full view of what is going on. When their time comes the birds are taken from the crates and shackled by their legs to the conveyor belt. This moves slowly towards the slaughterman. It is considered convenient to suspend the chickens in this way as it concentrates the blood in their heads and makes for more rapid bleeding when their throats are finally cut. Almost every stage of the 'processing' can be done while they are on the conveyor belt. The live birds can hang for anything up to five minutes before they are killed. *Photograph: by courtesy of 'The People'.*

**Figure 9.** Photograph: by courtesy of 'The People'.

**Figure 10.** Some packing stations have stunners to render the birds unconscious before they reach the slaughterman, but vast numbers go through, as in the foregoing photograph, to have their throats cut in full consciousness. They then flap their blood out in a 'bleeding tunnel' at the other end of which is a scalding tank. It is estimated that of the birds which have their throats cut in full consciousness, two out of every five go into the scalding tank alive. A leading veterinary surgeon has stated that in his opinion jugular severance without prior stunning is grossly inhumane as the birds are obviously in great pain for some appreciable time. The difficulty here is that invention of suitable stunners has not kept up with the ever increasing speed of the conveyor belt. Up to 4,500 birds are slaughtered each hour in large packing stations and it has not been felt incumbent on the industry to increase its speed only in relation to the availability of stunners. Legislation is being fought for and is sorely needed here. *Photograph: by courtesy of 'The Observer'.*

**Figure 11.** Photograph: by courtesy of C.A.A.C.A.

**Figure 12.** Calves for veal production are taken from the unwanted 'bobby' calves sold in markets. But their miseries are not then over, rather do they intensify. For the veal calf's life is one of deprivation from first to last. *Photographs: above by courtesy of 'The Daily Mirror'; (below) Dex Harrison.*

**Figure 13.** For the first two weeks of their lives the calves are allowed a little straw to lie on for warmth and comfort but after that age they are allowed only the bare slats as they might satisfy their craving for roughage and for iron by chewing the straw. These photographs show quite clearly the shortness of the tether.

**Figure 14.** Feeding-time. The end calf takes a moment off to lick the door handle. These calves will lick any metal to satisfy their needs. It is worth noting that immobilised and uncomfortable as they are, veal calves have to bear the full brunt of a heat wave without even being allowed to moisten their mouths. They must be kept thirsty so that they will drink the abnormally high amount of milk substitute offered to them at each feed. This milk substitute contains a very high proportion of fat to help them to put on weight. During the last year or two the tendency has been for broiler beef calves to be housed in much the same way as veal calves. Many are also deprived of roughage for which, like the veal calves, they develop a craving, being fed on concentrates only. Two points are worth noting: firstly, that veal calves are slaughtered at 3 months whereas beef calves are slaughtered at 12 months; and finally, that at the time of going to press there are 20,000 veal calves reared each year but hundreds of thousands of beef calves. *Photograph: Dex Harrison.*

**Figure 15.** The veal producer's ultimate aim is to fulfil the snob demand made in all innocence and ignorance by the public, for a white meat. To this end the calf is immobilised by a collar round its neck secured by a very short tether to two bars, enabling it to slide up and down but not permitting it any other movement. It is on slats often in near darkness, sometimes in a crate. It is then denied the feeling of being part of a herd which is so important to the well-being of a calf. Some calves have their heads caught permanently between upright bars.

Another serious deprivation is in their food. They are fed solely on a milk substitute and not allowed any of the roughage for which a ruminant craves, moreover the milk substitute is seriously low in iron, vitamin A and other ingredients, and is designed to keep the calf anaemic. The calves in this photograph only have their shutters lifted at feeding time, otherwise they are completely enclosed. Note that even in crates these calves are still tethered. *Photograph: by courtesy of 'The Daily Telegraph'*.

**Figure 16.** Photograph: by courtesy of *'Farmer and Stockbreeder'*.

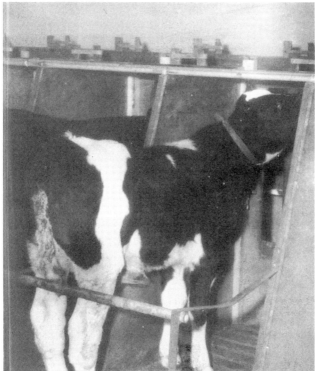

**Figure 17.** These two details show the space and conditions allotted the veal calf on an average farm. Note the tether at its head and the bar behind its legs, both aimed at immobility. The tail is matted with dung and the calf is powerless to do anything about the flies which torment it. Note the swollen knee joints caused by the deficiency diet and by the strain of constant balancing on slats. In heavier calves the hoof itself is sometimes distorted. *Photographs: Dex Harrison.*

**Figure 18.** One of the best and one of the worst veal farms I have visited. The animals above are untethered and on straw which is sprayed with a disinfectant to prevent browsing. The sun streams in through open shutters. Their only deprivation is in their food. Those below have their heads caught permanently between upright bars. Note how dirty the whole of the hind quarters are of the calf nearest to the camera. An attempt has been made in this unit to shade out the sun. The restlessness of the calves made photography difficult. *Photographs: Dex Harrison.*

**Figure 19.** Housing is not always ideal for the calf even on the traditional farm, but at least this one is on a long tether and lying comfortably on straw.

Note the rough rope used to tether the veal calves below. In this photograph, although superficially the conditions appear to be the same as those on the facing page, these calves, though tethered, do not normally have their heads caught between the bars except at feeding-time. The second calf could not stand having its head caught at any time. *Photographs: Dex Harrison.*

**Figure 20.** Battery cages are mostly put into controlled environment houses. Whereas with other animals geneticists aim through highly selective breeding to produce strains which convert food into flesh most readily, with hens they breed for rapid egg production, and any hen not achieving sufficiently high numbers is replaced by another hen. A battery house is an egg factory and is automated and run like one. Endless experiments are carried out by research workers: for example it was found that removal of combs and wattles resulted in less food being eaten and more eggs laid, that more lighting, or less, would produce the same end, that a yellow dye introduced into the feed would produce the golden yolk which the housewife associates with quality. A serious problem to the health authorities in battery houses is that the droppings trays running below each tier of cages provide a perfect breeding ground for flies. On a solid floor the hens themselves keep this under control by eating the larvae but here the flies breed in peace and have become resistant to nearly all fly sprays. *Photograph: by courtesy of Sterling Poultry Products Ltd.*

**Figure 21.** Far less food produced by natural methods is needed than of a cheaper forced product, and in the long run results in better consumer health. There is a danger that the momentum with which experiments in the industry sweep on precludes sufficient study of all the side effects of the research. Is there sufficient awareness, for example, of the final effect on man of all the antibiotics, hormones, tranquillisers, insecticides and growth stimulants used so lavishly in the industry? Serious concern is felt by some scientists and this should surely make us pause and consider where it is all leading and whether we are not handing on a poor heritage to our children in the name of efficiency and progress. *Photograph: A.C. Moore.*

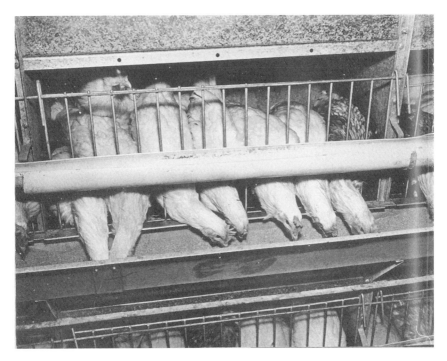

**Figure 22.** Some pullets are reared in battery cages and never know freedom. The hen's sole function in life is to lay as many eggs as possible during the year of laying life she is permitted before joining the broilers on the slaughter line. At first only one bird per cage was considered adequate, then two were put in each cage and survived, so now three birds to a cage is the accepted thing. You can see how much space for manoeuvring this allows the bird! The height of some cages has now been lowered so that the hen has to put her neck out of the cage to stretch it. She stands on wire mesh which has a slope of one in five to allow her eggs to roll away. Droppings are caught on a belt below. The cage is served by automatic food and water troughs. *Photographs: (above) by courtesy of 'Farmer's Weekly'; (facing) Kenneth Oldroyd.*

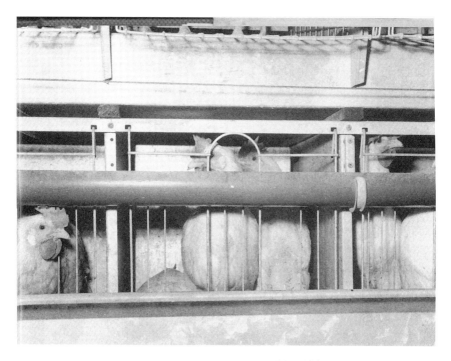

**Figure 23.** Hm-m-m … another sunny day.

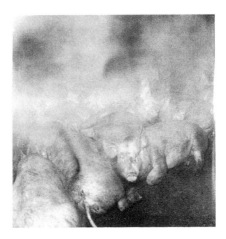

**Figure 24.** Phew-w ... w! *Photographs by courtesy of 'Farmer's Weekly'.*

# The Basis of Quality

There has probably never been a time of greater food consciousness than today. We have available in our shops a wide variety of foods from all over the world. They are prettily packaged and easy to prepare, some even ready-cooked, complete with recipes as to how and with what they are best served. How easy it is for the busy housewife to prepare a sophisticated and tasty looking meal.

But how much does the average housewife (or her husband for that matter) know about food, how it is grown, how it is processed, how much nutritional value different foods have? She imagines she knows quite a bit about calories, proteins and vitamins, because she has read so many articles on dieting, especially on slimming. She probably gives the matter a good deal of thought when she goes shopping. But how much is she really ever told? How often is she told the difference between white bread and brown bread, white sugar and brown sugar? Does she really know that one is a devitalised, demineralised, empty version of the other? That one slice of stone-ground wholemeal bread is worth more in food value than five of white? Does the poor mother with a large family of mouths to feed, or the old age pensioner on his pittance, ever stop to consider what poor value they are buying when using white bread as their cheap filler-upper?

How many people, in fact, ever stop to think that the food we eat today becomes our bodies of tomorrow – 'we are what we eat' in the words of the German proverb?

Food is the basis of our lives. Without adequate amounts of food we die and without adequate amounts of the right foods, we are inefficient,

© J. Harrison and J. Wilson 2013. *Animal Machines* (Ruth Harrison)

unwell, unable to enjoy life to the full, to work properly, or to think wisely or well. Sir Robert McCarrison, surgeon and nutritionist, explains the processes of nutrition in his book *Nutrition and Health*:

> The processes involved in the function of nutrition are mastication, deglutition, digestion, absorption, circulation, assimilation and excretion; the last including perspiration, exhalation, urinary excretion and defaecation. There are thus three stages in nutrition: the first, effected in and by the alimentry tract; the second, in or by the cells composing the body; and the third by the organs of excretory functions, skin, lungs, kidneys and bowel. It is of the utmost importance to realise that not only is the activity proper to the function of nutrition dependent on the efficient performance of all these acts, but their efficient performance is dependent on the adequate nourishment and functional efficiency of the organs and tissues performing them.
>
> At this point, and to maintain the sequence of our story, reference might be made to the implications of these acts: mastication, digestion, absorption, assimilation, and so on. But it may be enough to remind you that they include the ordered operation of involuntary muscular action, the production of various digestive and other juices, the elaboration of ferments, or enzymes, and of catalytic agents needed for the speeding-up of chemical processes, the production of blood-forming substances, the interchange of body fluids, the transport of nutrients to the remotest recesses of the body, the removal of end-products of chemical action and waste products from the body, and many other vital processes; all of which are influenced favourably or unfavourably by the constitution of the food. The alimentary tract and the organs (including the teeth) associated with it are of particular importance in this connection. They form a highly specialised mechanism designed for the nourishment of the body. The efficiency of the function of nutrition depends primarily on the functional efficiency of this mechanism and this, in its turn, on the constitution of the food....

It becomes increasingly obvious that we should all have a basically sound understanding of what we are eating and the way in which it is grown and processed. This is entirely a matter of common sense and does not mean that we need in any way become faddists.

The preoccupation of medicine has always been in curing sickness rather than establishing health. Dr Wrench in *The Wheel of Health* describes his medical training:

> We started from our knowledge of the dead, from which we interpreted the manifestations, slight or severe, of threatened death, which is a disease. Through these various manifestations, which fattened our text-books, we approached health. By the time, however, we reached real health, like that of the keen times of public school, the studies were dropped. Their human representatives, the patients, were now well, and neither we nor our educators were any longer concerned with them. We made no studies of the healthy – only the sick. Disease was the reason for our specialised existences.

There was also a great abundance of it. Between its abundance and its need to ourselves its inevitability was taken for granted. ...

Dr Wrench, and a handful of other far-sighted doctors, were not entirely happy with the direction of their training. They believed that it would have been more useful to start at the other end of the story, to study health and what goes to make a really healthy person, and consequently the results of deviation from the course of a healthy life. This, they felt, and still feel, would take us nearer to a basically healthy community than the fragmentary thinking of doctors today. They were considered cranks in their profession, as are most people who do not conform to mass ideas, but they persevered, and despite lack of financial backing for their research, carried it through to interesting and conclusive results. Many of them have written books on their theses and these (listed in the bibliography) make fascinating and completely convincing reading.

The first thing they had to do was to seek out those communities enjoying perfect health and make a study of them to discover what it was in their way of life which differed from our own, so that they enjoyed health while we were lucky if we were free of sickness. We have seen from McCarrison that health is a positive factor in itself, involving the perfect working of every function of the body, and not the usual medical interpretation of mere freedom from sickness.

Dr Weston Price, an American dental surgeon, decided with his wife, to spend their vacations studying communities where health and physique were outstanding and his detailed accounts of some fifty such tribes scattered all over the world makes excellent reading. His findings, in a nutshell, were as follows (*Nutrition and Physical Degeneration*):

> After spending several years approaching this problem by both clinical and laboratory research methods, I interpreted the accumulating evidence as strongly indicating the absence of some essential factors from our modern program, rather than the presence of injurious factors. This immediately indicated the need for obtaining controls. To accomplish this it became necessary to locate immune groups which were found readily as isolated remnants of primitive racial stocks in different parts of the world. A critical examination of these groups revealed a high immunity to many of our serious affections so long as they were sufficiently isolated from our modern civilisation and living in accordance with the nutritional programs which were directed by the accumulated wisdom of the group. In every instance where individuals of the same racial stocks who had lost this isolation and who had adopted the foods and food habits of our modern civilisation were examined, there was an early loss of the high immunity characteristics of the isolated group. These studies have included a chemical analysis of foods of the isolated groups and also of the displacing foods of our modern civilisation.

These isolated peoples from all parts of the world varied in religion, climate, diet, environment. What they had in common – the only thing they had in common – was that their food was cultivated from a fertile soil and eaten whole and fresh from the source. This is what gave them their perfect physique and when civilisation encroached on their lives with its bounty of 'civilised' white sugar, white flour and processed foods, the difference in health and vigour was soon apparent.

Price traces the importance of nutrition much further than we have space to discuss here, for he came to believe that a really healthy nation suffers very little from afflictions of hatred and unhealthy minds but lives peaceably together recognising each his place in the life of the community. McCarrison bore this theory out in his famous experiment on rats at Coonoor.

The findings of these doctors, and their appeal for the understanding of quality in food and its direct relationship to health and consequently to the whole basis of life, fired the local Medical and Panel Committees of the County Palatine of Cheshire, committees covering the six hundred family doctors of Cheshire, to write their indictment of the National Health Insurance Act in as far as they did not feel that it fully comprehended this more profound interpretation of health.

It is a remarkable document, issued more than twenty years ago, but as searching in its questioning and apposite in its conclusions as if it were written today. I do not know what response this testament had in its day, apart from a strong local influence on farmers and inhabitants, but these doctors must feel the lack of fundamental thought at individual and national level even more acutely today than they did then.

Here is an extract:

> How far has the Act fulfilled the object announced in its title – 'The Prevention and Cure of Sickness'?
>
> Of the second item we can speak with confidence. If 'postponement of the event of Death' be evidence of cure, that object has been achieved: the greater expectation of life which is shown by the figures of the Registrar General is attributable to several factors; but certainly not least to the services of the panel.
>
> The fall in fatality is all the more notable in view of the rise in sickness. Year by year doctors have been consulted by their patients more and more often, and the claims on the benefit funds of Societies have tended to rise.
>
> Of the first item, 'the Prevention … of Sickness' it is not possible to say that the promise of the Bill has been fulfilled.
>
> Though to the sick man the doctor may point out the causes of his sickness, his present necessity is paramount and the moment is seldom opportune, even if not altogether too late for any essay in preventive medicine. On that first and major count the Act has done nothing.
>
> We feel that the fact should be faced.

Our daily work brings us repeatedly to the same point: 'this illness results from a lifetime of wrong nutrition! ...'

It seems to us that the master key which admits to the practical application of this knowledge as a whole has been supplied by Sir Robert McCarrison.

His experiments afford convincing proof of the effects of food and guidance in the application of the knowledge acquired.

In describing his experiments, which were made in India, he mentions first the many different races of which the population, 350 million, is composed.

'Each race has its own national diet. Now the most striking thing about these races is the way in which their physique differs. Some are of splendid physique, some are of poor physique, and some are of middling physique. Why is there this difference between them? There are, of course, a number of possible causes: heredity, climate, peculiar religious and other customs and endemic diseases. But in studying the matter it became evident that these were not principal causes. The principal cause appeared to be food. For instance, there were races of which different sections came under all these influences but whose food differed. Their physique differed and the only thing that could have caused it to differ appeared to be food. The question then was how to prove that the difference in physique of different Indian races was due to food. In order to answer it I carried out an experiment on white rats to see what effect the diets of these different races would have upon them when all other things necessary for their proper nutrition were provided. The reasons for using rats in experiments of this kind are that they eat anything a man eats, they are easy to keep clean, they can be used in large numbers, their cages can be put out in the sun, the round of chemical changes on which their nutrition depends is similar to that in man, and, a year in the life of a rat is equivalent to about twenty-five years in the life of a human being. So that by using rats one gets results in a few months which it would take years to get in man. What I found in this experiment was that when young, growing rats of healthy stock were fed on diets similar to those of people whose physique was good the physique and health of the rats were good; when they were fed on the diet similar to those of people whose physique was bad the physique and health of the rats were bad; and when they were fed on diets similar to those of people whose physique was middling the physique and health of the rats were middling.'

A special group which he fed on the food of Travancore, in which there is a considerable portion of tapioca, disclosed a far higher percentage of gastric and duodenal ulcer cases than the other groups. This was informing as the people of Travancore suffer with peptic ulcers very much more commonly than the other peoples of India.

'Good or bad physique as the case might be was, therefore, due to good or bad diet, all other things being equal. Further, the best diet was one used by certain hardy, agile, vigorous and healthy races of Northern India.' (Note: the Hunza, Sikh and Pathan.) 'It was composed of freshly ground whole wheat flour made into cakes of unleavened bread, milk, and the products of milk (butter, curds, buttermilk), pulses (peas, beans, lentils), fresh green leaf vegetables, root vegetables (potatoes, carrots), and fruit, with meat occasionally.

'Now in my laboratory I kept a stock of several hundred rats for breeding purposes. They lived under perfect conditions; cleanliness, roomy cages, good bedding, abundant fresh water, fresh air and sunlight – all these things they had; and they were fed on a diet similar to that race whose physique was very good. They were kept in stock from birth up to the age of two years – a period equivalent to the first fifty years in the life of human beings. During this period no case of illness, occurred amongst them, no death from natural causes, no maternal mortality, no infantile mortality except for an occasional accidental death. In this sheltered stock good health was secured and disease prevented by the combination of six things: fresh air, pure water, cleanliness, sunlight, comfort and good food. Human beings cannot, of course, be so sheltered as these rats were, but the experiment shows how important these things are in maintaining health.

'The next step was to find out how much of this remarkably good health and freedom from disease was due to the good food: food consisting of whole wheat flour cakes, butter, milk, fresh green vegetables, sprouted pulses, carrots and occasionally meat with bone to keep the teeth in order. So I cut out the milk and milk products from their diet or reduced them to a minimum, as well as reducing the consumption of fresh vegetable foods while leaving all other conditions the same. What was the result? Lung diseases, stomach diseases, bowel diseases, kidney and bladder diseases made their appearance. It was apparent, therefore, that the good health depended on the good diet more than on anything else and that the diet was only health-promoting so long as it was consumed in its entirety, so long, in fact, as it contained enough milk, butter and fresh vegetables.

'Many more experiments were done which showed that when rats or other animals were fed on improperly constituted diets, such as are habitually used by some human beings, they developed many of the diseases from which these human beings tend to suffer: diseases of the bony framework of the body, of the skin covering it and of the membranes lining its cavities and passages; diseases of the glands whose products control its growth, regulate its processes and enable it to reproduce itself; diseases of those highly specialised mechanisms – the gastro-intestinal tract and lungs – designed for its nourishment; diseases of the nerves. All these were produced in animals under experimental conditions by feeding them on faulty human diets. Here is an example of such an experiment: two groups of young rats, of the same age, were confined in two large cages of the same size. Everything was the same for each group except food. One group was fed on a good diet, similar to that of a Northern Indian race whose physique and health were good, and of which the composition is given above. The other was fed on a diet in common use by many people in this country: a diet consisting of white bread and margarine, tinned meat, vegetables boiled with soda, cheap tinned jam, tea, sugar and a little milk: a diet which does not contain enough milk, milk products, green leaf vegetables and whole-meal bread for proper nutrition. This is what happened. The rats fed on the good diet grew well, there was little disease amongst them and they lived happily together. Those fed on the bad diet did not grow well, many became ill and they lived unhappily together; so much so that by the sixtieth day of the experiment the stronger ones among them began to kill and eat the weaker, so that I had to separate them. The diseases from which they suffered were of three chief kinds: diseases of the lungs, diseases of the stomach and intestines, and diseases of the nerves; diseases from which one in every three sick persons, among the insured classes, in England and Wales suffer.'

These researches were minutely made on a large scale and, but for the food, the conditions of each group were identical and ideal. Their results to our minds carry complete conviction – especially as those of us who have been able to profit by their lesson have been amazed at the benefit conferred upon patients who have adopted the revised dietary to which that lesson points.

It is far from the purpose of this statement to advocate a particular diet. The Esquimaux, on flesh, liver blubber and fish, the Hunza or Sikh, on wheaten chappattis, fruit, milk, sprouted legumes and a little meat; the islander of Tristan on his potatoes, sea-birds' eggs, fish and cabbage, are equally healthy and free from disease.

But there is some principle or quality in these diets which is absent from, or deficient in, the food of our people today. Our purpose is to point to this fact and to suggest the necessity of remedying the defect.

To descry some factors common to all these diets is difficult and an attempt to do so may be misleading since knowledge of what those factors are is still far from complete; but this at least may be said, that the food is, for the most part, fresh from its source, little altered by preparation and complete; and that in the case of those based on agriculture, the natural cycle:

$$\left.\begin{array}{l}\text{Animal and}\\\text{Vegetable}\\\text{waste}\end{array}\right\} - \text{Soil} - \text{Plant} - \text{Food} \left\{\begin{array}{l}\text{Animal} -\\ -\end{array}\right\} \text{Man is complete.}$$

No chemical or substitution stage intervenes.

Sir Albert Howard's work on the nutrition of plants, initiated at Indore and carried from India to many parts of the world, seems to constitute a natural link in this cycle.

He has shown that the ancient Chinese method of returning to the soil, after treatment, the whole of the animal and vegetable refuse which is produced in the activities of a community results in the health and productivity of crops and of the animals and men who feed thereon.

Though we bear no direct responsibility for such problems, yet the better manuring of the homeland so as to bring an ample succession of fresh food crops to the tables of our people, the arrest of the present exhaustion of the soil and the restoration and permanent maintenance of its fertility concerns us very closely. For nutrition and the quality of food are the paramount factors in fitness. No health campaign can succeed unless the materials of which the bodies are built are sound. At present they are not.

Probably half our work is wasted, since our patients are so fed from the cradle, indeed before the cradle, that they are certain contributions to a $C_3$ nation. Even our country people share the white bread, tinned salmon, dried milk regime. Against this the efforts of the doctor resemble those of Sisyphus.

This is our medical testament, given to all whom it may concern – and whom does it not concern?

We are not specialists, nor scientists, nor agriculturists. We represent the family doctors of a great county, the county, said Michael Drayton, of 'such soundly feed'; a county which gives its name to a cheese than which there is none better, though to most Englishmen, alas, only a name; a county where the best farming is still possible, which should minister to the needs of its own industrial areas and of a far wider circle.

We cannot do more than point to the means of health. Their production and supply is not our function. We are called upon to cure sickness. We conceive it to be our duty in the present state of knowledge to point out that much, perhaps most, of this sickness is preventable and would be prevented by the right feeding of our people.

This is the conclusion, that to maintain people in a state of wholeness, or health, in which every organ of the body is functioning correctly, we must follow the natural cycle of life, soil, plant, animal, man, soil, and as soon as we tamper with this cycle we lose some degree of health and immunity to disease.

It would be apposite to mention here an interesting long-term experiment being conducted by the Soil Association at their farm in Haughley, Suffolk, in an attempt to examine this thesis. The farm is divided into three sections to compare and contrast the nutritional effects, on successive generations of farm animals, of food grown from successive generations of crops, nurtured under different systems of soil treatment, on adjoining units of land of similar soil types, and under the same management. One unit was kept as a purely Organic section, with stock but without any help from outside food or fertilisers. The second unit, a Mixed section, has been run as a conventional farm with maximum advantage taken of fertilisers and so on. The third, the Stockless section, whilst run on conventional lines, carries no stock.

After twenty-five years a progress report has been written which suggests that proof is once more, under our very noses, being produced to bear out the ideas of the pioneers previously quoted. Let me quote the summary of the trends they are finding at Haughley:

The fertilisers, used to supplement the organic manuring on the Mixed section, have usually (though not always) resulted in the production of a somewhat higher bulk tonnage of both fodder and grain than that obtained from the Organic section, as well as lusher pastures. As a result the cattle and poultry on the Mixed section have usually received a 5 to 15 per cent higher allocation of winter rations than those on the Organic section.

In spite of this greater food intake on the Mixed section, milk production from 1956, when 2nd generation animals were coming in to the herds, has been consistently higher from the Organic section herd, whether measured as total milk produced, or production per cow, or production per acre. At the same time the cows on the Organic section have carried better condition, more 'bloom', and have shown quite clearly a greater contentment and placidity. *This more-milk-for-less-feed experience has been one of the most interesting farming findings to date*, and is particularly marked when milk is being produced from grazing only…. (The italics are mine.)

'The next most interesting observation,' the report tells us, 'has been the increasingly "self-supporting" nature of the Organic section crops. Those on the other two sections are definitely dependent on the artificial aids they receive. This was to be expected on the Stockless section; but it has been demonstrated, even on the Mixed section that, if fertiliser is omitted from even a small area of any field, the yield of that part of the crop drops well below the yield of the equivalent Organic crop. Conversely, the heaviest yielding fields on the Organic section are those that have been *longest* without fertilisers (35 years in some cases), suggesting that no depletion is taking place. The Organic section crops appear also to be less susceptible to insect pests, and rarely show any deficiency symptoms.'

Other trends during the ten year period that have been noticeable in the course of ordinary farming operations have included improved 'workability' under all weather conditions of the organic fields, and a marked sectional difference, observable only from the 3rd generation, in the proportion of small to large grain in cereals, percentages being of the order of 10 per cent 'smalls' on the Stockless crops, 7 per cent on the Mixed, and only 5 per cent on the Organic, as obtained by sieving for seed purposes.

The comparative health of the livestock of the two sections has naturally been watched with great interest. For the first few generations there was little or nothing to choose between them, both being high, but during the last year or two, indications of loss of stamina have made their appearance in the Mixed stock.

More milk for less feed, less bulk but richer crops, healthier and more contented stock, what a difference between this picture and that of our drug-drenched immobilised animals in their dark sheds. Research at Haughley is to continue and surely deserves recognition at national level. It is the only scientific experiment of its kind in the world and its future findings will undoubtedly prove immensely valuable.

I will examine in the next chapter how far we have moved away from this ideal in our modern conception of agriculture where the struggle for quantity production overrides all ideas of quality, and we will see whether this has any relationship to the ever expanding number of diseases with which medicine is now having to cope.

# Quantity versus Quality

Health is a positive quality in itself and does not simply denote an absence of illness. We have seen that those primitive communities who have achieved health have done so by eating food fresh from its source, unadulterated by processing, and equally important, from a source which, in its turn, is also healthy. Circumstances, and the wisdom of the tribe handed down from generation to generation, made this possible in small communities. When we turn to the vast populations of the world as a whole, and especially to the concentration of people in heavily populated areas, some modification from the ideal is inevitable.

Of environmental factors concerning health there have been both improvement and deterioration. Housing improves steadily and so do working conditions. Our basic standards of comfort and cleanliness get better all the time and we are making some headway in preventing the terrible pollution of air from industrial chimneys which has wrought such havoc on health. But against this must be set the increasing pace of life especially in towns, with its byproducts of overcrowding, rush-hour travel, queueing and so on. We suffer more and more from exhaust fumes of ever-growing numbers of cars, buses and heavy lorries. There is the vastly increased burden of noise. Air travel has brought its shattering of the peace to town and country alike.

All this is to some extent inevitable, but some of the greatest hazards besetting man today are those which he himself has needlessly made. His preoccupation with the idea that to prevent war he must make supplies of ever bigger nuclear bombs in the testing of which he releases greater

© J. Harrison and J. Wilson 2013. *Animal Machines* (Ruth Harrison)

and greater levels of radiation into the atmosphere, his even wider contamination of life with the reckless use of insecticides and pesticides, and his destruction of some of the basic nutrients of food through processing, are things which the individual must find it hard to credit in this age of enlightenment.

Of what profit is it to man that he earns more, lives better, has a wider culture, if he is unprepared to fight for the one gift also handed down to his children, without which all others are useless, the gift of health?

It is generally assumed that the expectancy of life has increased in the last fifty years. But this needs some qualification. It is the expectancy of life at birth which has improved, not that of old age, and this is because we have discovered wonderful new drugs with which to suppress the infectious diseases so mortal to children half a century ago. A child has a far greater chance of reaching the age of forty today, but after that age the difference in the expectancy of life is only marginal. That this should be so when new drugs and techniques are available to us unheard of even twenty years ago, when also basic factors like housing and sanitation are, greatly improved, indicates failure somewhere along the line.

This failure is all the more sad when we reflect that not only is illness not less today, but a basic degree of illness is accepted without question as a part of our lives. Fatigue, headaches, digestive troubles, constipation, are taken for granted, as part of modern living. An equal reflection that something is basically wrong is the steady increase in degenerative diseases – in diseases of the heart, ulcers, diabetes, bad teeth, cancer, and so on. We are taking these degenerative diseases into our lives as if they too are inevitable, and tend to shrug off statistics as being mere fodder for politicians until they come to bear on our intimate circle of acquaintance.

Who can fail to be shocked by the following statements by Rachel Carson (*Silent Spring*):

> … the American Cancer Society estimates that 45 million Americans now living will eventually develop cancer. This means that malignant disease will strike two out of three families.
>
> The situation with respect to children is even more deeply disturbing. A quarter-century ago, cancer in children was considered a medical rarity. *Today, more American school children die of cancer than from any other disease.* So serious has this situation become that Boston has established the first hospital in the United States devoted exclusively to the treatment of children with cancer. Twelve per cent of all deaths in children between the ages of one and fourteen are caused by cancer….

M. Berglas, of L'Institut Pasteur, Paris, says that in his opinion everyone will, before long, be threatened with death through cancer.

The chief preoccupation of the medical profession is with disease after it has arisen. The countless drugs available to the doctor are aimed at bringing relief, to suppress disease rather than to cure. The inherent limitations of this attitude are reflected in the rising cost of the Health

Service, which in 1951 was £486 million, in 1956 £624 million, and by 1961 had risen to £926 million.

As a sop to the farmer we are paying out some three to four hundred million pounds a year in subsidies, much of which goes in the production of food of such doubtful quality that it adds to the burden of ill health.

Can we, as individuals, do anything to break this vicious circle? The answer is, indeed yes. We can take a lead from the isolated tribes of whom we have read and try to feed our families on food as fresh and unadulterated as we can find. A really healthy person is far more able to withstand the hazards of our modern existence than one whose health and resistance has been undermined by poor food. That we are deliberately fostering the production of foods the nutrients of which are partially destroyed, and which have even traces of toxic substances in them, is not only shortsighted and bad economics, but little short of criminal.

Modern techniques of farming ensure the contamination of food right from its beginnings. Pesticides are often washed over seeds before planting and thereafter systematic chemical spraying aims at reducing loss through insect or parasite. Biological controls are made difficult owing to specialist farming. It is no longer considered economic to run a mixed farm, however good it might be for the fertility of the soil, and farmers are advised to concentrate for profit on one or two crops and on one or two species of livestock kept intensively.

This concentration of one crop over a vast area has enabled pests which thrive on that crop to gain such a hold that even persistent spraying with insecticides is proving ineffectual:

> Under primitive agricultural conditions the farmer had few insect problems. These arose with the intensification of agriculture – the devotion of immense acreages to a single crop. Such a system set the stage for explosive increases in specific insect populations. Single-crop farming does not take advantage of the principles by which nature works; it is agriculture as an engineer might conceive it to be. Nature has introduced great variety into the landscape, but man has displayed a passion for simplifying it. Thus he undoes the built-in checks and balances by which nature holds the species within bounds. One important natural check is a limit on the amount of suitable habitat for each species. Obviously then, an insect that lives on wheat can build up its population to much higher levels on a farm devoted to wheat than on one in which wheat is intermingled with other crops to which the insect is not adapted. (Rachel Carson, *Silent Spring*)

During the last twenty years over two hundred chemicals have been used in our war against insects, weeds and 'pests'. We are now discovering that rather than achieving their object the danger exists that the effect of these poisons may be boomeranging back on us.

> Some would-be architects of our future look towards a time when it will be possible to alter the human germ plasm by design. But we may easily be doing so now by inadvertence (remarks Rachel Carson), for many chemicals,

like radiation, bring about gene mutations. It is ironic to think that man might determine his own future by something so seemingly trivial as the choice of an insect spray.

Apart from this hazard attached to the actual growing of food, others exist in the additives aimed to protect food in storage and processing. Some of these are harmless, but others, such as some dyes, have been found to be carcinogenous. An equal danger is that some of the essential nutrients of the food are destroyed or drastically altered during processing.

The soil, the seed, the plant, and now the animal!

Intensification of livestock rearing has simply taken all these hazards a stage further, and in my opinion made the food a definite danger to consumers.

Veterinary surgeons must shudder when they enter the sheds where these animals are kept; all the basic concepts of health are so flagrantly broken.

The first big step away from health has been to divorce the animals from the soil.

They are put into sheds with concrete floors. If they are lucky, and make enough profit for the farmer that way, they are allowed litter to rest on, otherwise they must spend their lives on slats or wire mesh. This separation from the soil has also involved their food. Its own choice of food, as well as being pleasurable for the animal, can probably give it the basis of a far healthier diet than the monotony of compounded food, however carefully thought out this might be. Roy Bedichek, in *Adventures with a Naturalist*, points out that although synthetic vitamins are added to the feed by back-room boys, who take for granted that they know better than the animal itself what is needed to keep it healthy, there might yet be some item of knowledge which has escaped them:

> It is easy to say that the vitamin content or mineral content is just as high in the mash-fed as in the chicken that gets his natural food in a natural way, but the fact is that all the vitamins have not yet been isolated. Several others are suspected to exist. Until a complete vitamin list is available, how can one say, comparing two foods, that the vitamin content of one is just as high as that of the other? Moreover, research has not yet presented us with a complete survey of the results of every vitamin and the results of all permutations and combinations of vitamins. We are led to believe by some vitamin advertising that it's not necessary anyway to get your vitamins in food since you can get them much more easily, if more expensively, by taking pills.
>
> Let us rear a monument here to the Unknown Vitamin and place a wreath reverently at its base. Perhaps it lies concealed in the body of a grasshopper or other insect, so securely imprisoned that only the digestive apparatus of the chicken can make it available for men....

The next step has been to enclose the birds and animals, often in the dark or near-darkness, away from fresh air, sun and rain, and without the possibility of exercising their limbs in even the most rudimentary way.

Professor E. A. Muller, of the Max-Planck Institute in Germany, did an experiment to find the effect of two weeks immobility on a healthy medical student. This was to determine what would happen to an astronaut if he spent long in the confines of the space ship. *The Evening Standard* reported this experiment on 9th November 1962:

> Doctors fed, washed and carried him to and from the lavatory. Finally they examined him – and found he had lost 20 per cent of his strength … the man's muscles simply 'melted away', said the Professor, and he gained weight rapidly.

No doubt a cannibal would have found him very tender, but he was not considered to be healthy, and it was decided that the astronaut must be able to exercise all his muscles daily to keep fit.

It is easy to see the analogy of this experiment with our veal calves, baby beef calves and pigs, even with battery birds and broiler chickens. Their muscles also become flabby and they put on weight rapidly, *but they are not healthy*.

Is it surprising after all this that they have to be kept alive with drugs?

The deterioration of animal health is causing great concern to agriculturists and veterinary surgeons.

> Serious obstacle to extension of research into this complicated problem (says the report of the Animal Health Trust on plans for research into leucosis) has been the widespread existence of leucosis in poultry generally … (*Poultry World*, 17th May 1962)

and *Farmer and Stockbreeder* comments:

> … Dr Gordon (of Houghton Poultry Research Station) said there was every evidence that the chronic respiratory disease complex was now increasing and becoming a very serious problem.
>
> Leucosis was not only on the increase but occurring at an increasingly early age. In certain broiler flocks outbreaks had been of an explosive nature with an extremely high mortality rate. (4th September 1962)

And Mr David Bellis, B.O.C.M. Chief Pig Advisor, told a meeting that whereas dense stocking rates increased efficiency and profitability, they also increased disease risk: 'I may be pessimistic, but I believe that 90 per cent of our pigs suffer from some sort of clinical or sub-clinical disease,' he added (*Farmer and Stockbreeder* 14th March 1961).

'Turkey diseases are on the increase, becoming more virulent and more difficult to eradicate,' said Dr M. H. Fussell, poultry advisor to B.O.C.M. (*Farming Express*, 11th May 1961).

It has already been shown that health has no part in the veal programme, in that a degree of anaemia is vital to the production of 'white flesh'.

Unhealthy animals cannot make healthy food for humans. This is a statement with which, I think, few would argue. But this is not only unwholesome food, it is more than that, it is potentially dangerous. A race is developing between disease and the scientists creating new drugs to keep mortality down to levels where profit can still be made.

Drugs are used automatically, in small quantities in the compounded food to allow uninhibited growth, in larger quantities to suppress disease when it actually appears, and finally synthetic hormones are used for fattening. Traces of all these can be left in the carcase when the animal is slaughtered.

## Antibiotics

Dr Sainsbury, speaking at the British Broiler Growers Association Convention in October 1958 stated that he found broiler houses 'one of the finest mediums man has devised for the promotion of vice and disease'. I think it would be fair to add that modern factory farming methods could not be carried on without the use of antibiotics. We have seen that the calf, the pig, and the bird are allowed only the room needed for their actual bodies, and so close is their proximity to each other that it would be impossible for disease not to sweep through the sheds unless drastic methods of control were enforced.

'Why do we use oral antibiotics?,' asks the *Farmer and Stockbreeder* Vet., 'everyone knows and admits that they are only a substitute for good husbandry' (1st May 1962). The Ministry pamphlet on the subject informs farmers that 'weakly animals may benefit more than those which are more robust'.

Antibiotics suppress the bacteria in the intestinal tract of the animal and help to prevent their being handed on from animal to animal. They therefore serve two purposes: to stop widespread disease build up, and thereby to allow uninhibited growth of the animal.

Three antibiotics are now used as routine additives to the compounded feed of pigs, young poultry (other than breeding stock), and more recently, calves. These are penicillin, aureomycin and Terramycin, and their inclusion is recommended at the rate of 15 parts per million for pigs and 'rather less' for poultry. This, according to the Ministry, 'could lead to improved growth rates and to smaller amounts of food required per lb. of live weight gain'. And so, no doubt, to greater profits, which is one reason why their addition to feed has become routine. I would be the last person to wish to deny the farmer his profit, but have all the consequences of this routine oral injection of antibiotics been examined in sufficient detail, both for its long-term effects on animals and then on the last link in the chain, man himself?

In 1960 a Committee was set up jointly by the Agricultural Council and the Medical Research Council to study just these questions. In 1962 they produced their findings. They decided that there was little evidence in this country that antibiotics as growth additives were not being as successful as when first used – though they had to admit that in some other countries much more was having to be added to the feed to be effective. They had evidence also that strains of organisms could become resistant and harmful and, furthermore, could be passed from one animal to another so that therapeutic doses might in time become useless. *But they recommended the continued use of antibiotics on the grounds of economic gain.*

Even while they were thus allaying the misgivings of the farmer the hazard they foresaw had already come to pass.

> Examination of pigs and poultry kept on many different farms showed that the *B. coli* in the faeces of tetracycline-fed animals were predominantly tetracycline-resistant (observes H. William Smith of the Animal Health Trust), while those in the faeces of animals on farms where tetracycline feeding was not practised were predominantly tetracycline-sensitive. In some herds in which tetracycline feeding was just being introduced it was possible to trace the changes of the *B. coli* faetcal flora from tetracycline-sensitive to tetracycline-resistant. Smith notes that similar changes could be traced in the case of *Clostridium perfigens*. (Lewis Herber: *Our Synthetic Environment*)

The *Farmer and Stockbreeder* Vet. discussed this aspect of drug usage in an article, 1st May 1962, in which he said:

> Time and again I've seen strong porkers, bacon pigs, and especially sows, develop *E. coli* infections which have not responded to antibiotic treatment simply because the germ or germs concerned have built up a powerful resistance against that particular antibiotic through swallowing it in small quantities in the food over a lengthy period....
>
> Again and again I've seen litters a few days old dying off from virulent resistant *coli* picked up from the sow … when I have typed the bugs from the dead piglets and done sensitivity tests I have found the germs resistant to as many as three or four of our most potent antibiotics.
>
> Many times I've had to treat *E. coli* mastitis and metritis in sows by hit and miss techniques – because the causal bugs have acquired powerful drug resistance. In such cases, before an effective treatment is discovered the udder is ruined and the breeding life of the sow terminated....

This veterinary surgeon goes on to point out that wasteful of capital as this is the situation in Ireland and America is even worse. There *E. coli* and other bugs 'defy the lot' and farmers and veterinary surgeons have had to go back to 'aspirin, prayer and bottles of black magic'.

The danger has also arisen of pneumonia in pigs. The bacteria causing this have also acquired resistance and acute pneumonia is causing remorseless mortality. Resistance is also building up in the germ of swine erysipelas, and of poultry the Vet. makes the comment: 'I would say we are teetering on the brink of an abyss … surely and gradually the poultry bugs are becoming conditioned to the drugs and disaster cannot be so very far off.'

Might I comment that this picture does not seem so very like an economic gain to the farmers – rather perhaps an economic gamble.

And now what of the hazard to man?

The first obvious danger lies in man himself building up a resistance to antibiotics from continual ingestion of small, even infinitesimal amounts in his food and also in developing an allergy to these drugs by the same insidious means.

Dr Bicknell has found no figures as to how much antibiotic remains in the flesh of calves and pigs after slaughter, nor any evidence as to whether the feeding of tranquillisers to young cattle to increase their rate of growth leaves enough of 'these mischievous drugs in their meat to affect man'. With regard to hens, however, he quotes Frye et al. in this manner:

> ... After hens were fed for three weeks on food containing 1,000 parts per million of aureomycin or Terramycin or bacitracin the amounts found, respectively, in the meat were, in mg. per pound, 0·1 to 0·2, 0·05 to 0·1 and 0·07 to 0·09. Eggs from hens fed on 200 parts per million of aureomycin each contained about 0·01 mg. of the antibiotic. (*Chemicals In Food*)

Clearly the above figures represent intense dosage and the proportion remaining in the flesh is seemingly minute. The Ministry Committee also found only minute traces in the carcases of antibiotic fed animals.

As a layman I am not competent to judge what constitutes a minute dose of these drugs but Lewis Herber tells us (*Our Synthetic Environment*):

> The sensitivity of many persons to penicillin has reached a point where a therapeutic dose of the drug can be expected to claim their lives.
> Such individuals are increasing in number with every passing year.

He then goes on to quote experiments carried out by doctors who had patients, one 'with a reaction on passive transfer to 0·00001 unit of penicillin' and another whose patient went 'into shock when skin tested with 0·000003 unit of penicillin'. Are the traces in food in fact so minute that, in view of the above, they can be lightly dismissed?

Although only three drugs are permitted by law to be administered to livestock as a normal feed additive, there is no such limitation on the administration of drugs to animals to combat disease. They may be dosed indiscriminately with any available drug for this purpose. Do we know sufficient about these potent drugs to risk the hazard that some residue, however slight, will remain in the flesh or egg we then consume? Is this a risk that we should reasonably be asked to take, to take blindly and in complete ignorance that any risk is involved?

Then there are the dangers associated with resistant bacteria in animals: firstly that the bacteria can be passed direct from animals to man, and secondly, that they can be passed on in the food so produced.

To quote Bicknell again, in Australia

> ... a new strain of resistant staphylococci has spread in the general community, causing septicaemia with over a 40 per cent mortality ... it would seem most probable that the infection has spread from antibiotic treated animals to farm workers to everyone. (*Chemicals in Food*)

A major problem in hospitals, according to Bicknell, is the fatal infections of the bowel with antibiotic-resistant staphylococci. Young children get diarrhoea from *E. coli* which should respond to antibiotics but here also resistant strains have been found which could be linked with those so common on farms. He goes on to say

> Fungal infections in man may occur in the brain, lungs, gut, kidneys, skin, etc., and are now increasingly common since the antibiotics kill the bacteria which normally inhibit the growth of fungi. Most of these fungal infections have been fatal until now, or extremely unpleasant ... now there is some hope that the new antifungal antibiotics will cure such infections, but this hope is dimmed by the fact that farmers have forestalled doctors in the use of these antibiotics and presumably have already started to build up strains of resistant fungi, since resistance is rapidly developed.

The development of fungus disease is already much on the increase and causing some concern amongst turkey rearers.

Need I elaborate further?

Antibiotics taken in food might alter a person's body flora, warns Professor Frazer of Birmingham University (Public Health Conference Report 1962), giving rise to gastro-intestinal disturbances and other difficulties.

'Antibiotics are dangerous food additives,' says Lewis Herber.

'Let us abandon oral antibiotics before it is too late,' says the *Farmer and Stockbreeder* Vet.

And indeed it is ironic to think that while medical authorities are steadily urging that antibiotics be used only with great discrimination on the grounds of dangerous resistance building up, the agricultural authorities are encouraging ever wider use. Perhaps these two should get together some time to discuss the matter, before it is indeed too late.


## Other Additives

Other anti-infective agents are also fed regularly to poultry, such as the coccidiostats nicarbazin, sulphanilamido-quinoxaline and the nitrofurans. The last of these, says Dr Bicknell, 'often causes dermatitis in man when applied to the skin, but little is known about the general effects of coccidiostats when they are eaten frequently by man in very small amounts'.

A routine addition in turkey feed to prevent blackhead is acinitrazole, which is mildly toxic to man as a trichomonacide.

# Arsenic

'*Arsenic in all its forms* should be banned from use in agriculture and horticulture but, very shockingly, there seems to be no hope of this,' says Dr Bicknell, and Rachel Carson points out that arsenic 'was the first recognised elementary carcinogen (or cancer-causing substance),' and says that 'the association between arsenic and cancer in man and animals is historic'.

They were both referring to arsenic used as sprays in agriculture but it has also been used as a growth stimulant both here and in America.

> ... the F.D.A. has already discovered violations of its regulations on feed supplemented with arsenic compounds, another widely used promotor of growth (writes Lewis Herber). 'A very preliminary survey made some time ago by the Food and Drug Administration indicated that some poultry raisers are not withholding arsenic-containing feeds from their flocks 5 days before slaughter' (complained Arthur S. Flemming, former Secretary of Health, Education and Welfare), 'and we have information that in some parts of the country, hog-raisers maintain their animals on arsenic-containing feed within the five-day period that the arsenic is supposed to be withheld'. (*Our Synthetic Environment*)

Anyone familiar with the notorious arsenic poison cases of past decades would raise an eyebrow at the regulation five-day period quoted above. These famous poisoners operated over a period of months, slowly building up a lethal dose, and it was frequently months after the body had been laid to rest that it was exhumed and successfully tested for arsenic.

I have not heard of arsenical compounds being used as growth additives for pigs in this country, but they have certainly been used for poultry, and to levels causing concern to some authorities. In November 1961 the Wiltshire Weights and Measures Department reported (*Bristol Evening World*):

> For some time past the addition of arsenical growth stimulants to compound feeding stuffs designed for poultry feeding had given us some concern, particularly in the case of broilers.

They made four tests, three on birds from broiler houses and the fourth on the deep litter from a broiler house, which was then used as a fertiliser. They found that the birds stored arsenic in their livers so that whereas, for example, one lot of feeding stuff contained only 0·004 per cent of an arsenical additive, the livers from the birds fed on it contained 0·2 parts per million and 1·6 parts, thereby exceeding the permissible limit. Where the deep litter had been used as a fertiliser it was indicated that 'there was a small take-up' of arsenic in the grass.

The Advisory Committee on Poisonous Substances in Agriculture and Food were informed and *advised* feeding stuff manufacturers not to incorporate arsenical growth promotors in their compounds. 'This may be the end of the use of arsenicals in this country,' the Report adds hopefully, 'but we shall endeavour to keep an eye on the position.' I wonder if it was the end?

Incidentally, I have a letter from the Ministry of Agriculture informing me that '… it is possible that pâtés derived from the livers of broiler chickens which are currently on sale may be mistaken by some consumers for pâté de foie gras.' Next time you look along the row of pâtés in your delicatessen remember the arsenic which may, or may not still be there, and all the other chemicals and toxic substances liable to be stored in the livers of broiler chickens, and see whether you can persuade your delicatessen to define which are the broiler pâtés.

## Dyes

Other additives to feed include certain dyes, used to enhance the appearance. Every attempt, for example, is made to remedy the defects of the battery egg. To counteract the pale yolk, dried grass, or a yellow dye, is fed to the hens and makes for the rich golden yolk associated by the housewife with the quality of the free range egg. But the safety of this yellow dye is questioned by Dr Bicknell:

> … each of us is born with a capital sum which we can spend on neutralising cancer-causing chemicals. Each time we eat such a chemical we spend some of our irreplaceable capital: when the capital is spent we die of cancer.
>
> Therefore the individual cannot afford to eat anything which he thinks is cancer-causing. He must fight to leave himself all the margin of safety he can, so that he has used up none of his reserves before he is faced with some cancer-causing chemical which he will probably neither know exists nor be able to avoid, such as, possibly, some detergents and some eggs. Eggs are not 'food' until they have been laid by the hen. Therefore the pallid yolks of commercial battery eggs can legally be coloured with any yellow dye, however dangerous, if, being fed to the hen, it is excreted into the yolk. Thus deluding the public by providing battery eggs with yolks dyed to the golden yellow of the yolks of 'farm' eggs is a dangerous swindle and should be banned.

## Hormones

The following synthetic hormones are on the agreed list of permissible additives for some livestock: stilboestrol, hexoestrol, dienestrol, dianisyl hexane, dianisyl hexene, dianisyl hexadiene, dienoestrol diacetate and thyroid stimulants.

Dr Bicknell points out that

> … at least one form of cancer of the breast… is caused by or is dependent for its development and growth on oestrogens, that is on the

female sex hormones … in order to remove the stimulating effect of the oestrogens on cancer of the breast, such cancer is now often treated with surgery and X-rays combined with surgical removal of the woman's ovaries, that is, removal of the glands which make oestrogens.

This is followed by systematic removal of all organs which may make oestrogens. He goes on to say:

> Therefore it would seem essential that no external source of oestrogens should be given to women who are under treatment for cancer of the breast, or indeed, to any woman. *But beef, mutton and, especially, poultry are increasingly contaminated with artificial oestrogens* which have been injected into the animals in order to fatten them or increase their rate of growth … no woman should eat commercially bred chickens or meat that she knows have been treated with oestrogens.
>
> In defence of the treatment of animals with synthetic oestrogens, it has been said firstly that no more oestrogen is injected into an animal than it may make itself. But the *natural oestrogens present in the flesh of animals are destroyed during digestion in the human stomach while synthetic oestrogens are not thus destroyed but are absorbed into the body.*
>
> … Other cancers besides cancer of the breast are caused by oestrogens: leukaemia or cancer of the blood has probably been caused in man, and in animals cancer of the kidney, bladder, testis and uterus, and leukaemia.
>
> Oestrogen or hormone-treated animals should never be sold for human food. Until they are banned by law, the housewife should ask her butcher about his meat and poultry and refuse any which he thinks may have come from hormone-treated animals. (The italics are mine.)

Stilbestrol has been widely used in America as a synthetic hormone injected into the necks of chickens. It was considered safe on the assumption that the heads and necks of chickens were always discarded. This assumption overlooked the fact that many housewives were found to make stew from necks of chickens, and even heads. It was also discovered that whereas only one pellet was recommended for use, poultry farmers in their enthusiasm sometimes inserted two or even more. Leonard Wickenden quotes a report of an investigation made by the Food and Drug Administration in which they found,

> In one lot of 200 birds, 180 contained partially unabsorbed pellets. The residual stilbestrol per bird ranged in quantity all the way from 3 milligrams to 24 milligrams. These figures are significant in view of the testimony of one witness who stated that the recommended dose of stilbestrol, for therapeutic effects on human beings, ranged from 1 to 5 milligrams daily.
>
> When reading a report such as the above, the question which undoubtedly arises in many people's minds is what is the effect on a human being if he, or she, happens to absorb some of the stilbestrol when eating a treated chicken. On that question, a storm of argument raged before the Committee. Scientists representing the manufacturers or the poultry interests

protested that danger to consumers was virtually non-existent; disinterested scientists of high standing warned that the danger was real and grave. An edge was given to these warnings by the fact that groups of mink farmers who had purchased offal containing chicken heads to feed their minks were dismayed to find that the males became sterile, so that the breeding programme was disrupted. The general health of the animals was also affected, one witness stating that 'the animals that got stilbestrol were the poorest mink I have ever seen'.

It is interesting to note that in a similar case in Scotland damages were awarded to a firm of mink breeders, when on being fed poultry offal 'the mink herd became sterile. None of them bred, many of them died, and the rest were disposed of' (*Poultry World*, 21st February 1963).

Wickenden goes on to report that evidence was given to the Delaney Committee showing that stilbestrol, besides producing sterility, can, according to Dr Robert Enders, Professor of Zoology at Swarthmore College, if given in sufficient quantities, 'depress the growth of children, cause cystic ovaries, cystic breasts, cystic kidneys and suppress ovulation. …' Dr Enders goes on to point out, however, that far larger amounts would be necessary for this than could be ingested by eating the flesh of the chickens. He gave a warning, however, that stilbestrol becomes concentrated in the livers of chickens and these are often sold separately. He testified moreover

> that two hundredths of a milligram given over a period of time will kill an animal, whereas as much as two-milligram doses given over the same period to an animal of equal weight will not kill him. 'To me,' said Dr Enders, 'that can mean only one thing – that small doses are much more toxic than large doses.'

Further evidence to this effect was supplied by Dr Carl G. Hartman, of the Ortho Research Foundation, who

> … stated that it was widely believed in the medical profession that estrogen is a means of stimulating cancer. Estrogen given to a rat produced cancer in 3 months. He said: 'We find if you give a little and then stop a while, then give a little more and stop a while again, and then give a little more, it is more effective than giving it continuously. Also, you give a little – not much. If you give a great deal, you may not get any cancer. In fact, if you give a large dose, you inhibit cancer.'

Several specialists gave evidence that hormones should only be used under specialist advice, so grave were the consequences of indiscriminate use.

> Clearly (Wickenden goes on), the consumer faces a risk of damage to his health and, possibly, to his bodily functions. Does stilbestrol offer him compensation in the form of more nourishing meat at lower cost? If he has to carry most of the risk, should he not, in common justice, receive the larger share of the gain?

Dr Enders is quite emphatic that the meat is not more nutritious.

> On the economic side, I agree with those endocrinologists who say that the use of the drug to fatten poultry is an economic fraud. Chicken feed is not saved; it is merely turned into fat instead of protein. Fat is abundant in the American diet, so more is undesirable. Protein is what one wants from poultry. By their own admission, it is the improvement in appearance and increase in fat that makes it more profitable for the poultryman to use the drug.

Wickenden takes the argument further:

> Quoting the work of one of his students, Dr Enders said it had been found that the gain in weight, following the use of stilbestrol, was chiefly due to greater water retention in the fat. The fat of a treated fowl 'contains a great deal more water than that of a normal bird'. Asked if it made the bird heavier, he said, 'It makes the bird a great deal heavier. I would say you would get 5 lb., where you would get 4 lb., under normal conditions.' This added weight, he said, would be fat and water. He denied that the treatment would produce more breast, but said that 'the skin of the animal becomes very nice and smooth because there is water and fat under it. It increases the attractiveness very much.

Summing up the evidence given before the Delaney Committee he comments:

> … one learned specialist after another has given us his solemn warning. They have stated, in all earnestness, that stilboestrol is a dangerous compound, far-reaching in its effects. They have told us that it should be dispensed only under prescription, that it is not safe for every Tom, Dick and Harry to use, that it is biological dynamite. To these statements they have added that it is an 'economic fraud', producing increased weight without corresponding increased nourishment. They have told us that minute doses, repeatedly given, are more potentially dangerous than single large doses.
>
> Is it not the repeated, small dose which the American people are now being offered? If we take a little stilboestrol with chicken, a little with beef, a little with perhaps mutton and pork, shall we not be facing the exact conditions described by Dr Enders and Dr Hartman which, they said, were 'much more toxic' than those produced by a single large dose?
>
> All such warnings are swept aside; we ignore them as though they had never been given…. We are ready to accept, without surprise, almost without indignation, the fact that those responsible for producing our food will, with little hesitation, urge the adoption of any procedure that will bring, even temporarily, quicker or easier profits. The consumer is almost forgotten. A suggestion that we should wait a few months, or even a few years, to see whether the health of human beings will be affected, is considered laughable.
>
> There is no final proof that cystic ovaries, or cystic breasts or cystic kidneys will result from a widespread use of stilbestrol. There is only a

possibility. There is no final proof that human sterility will follow; again, only a possibility. There is no final proof that subtle and profound changes will occur in our bodies, bringing sorrow and ruin to nobody knows how many lives. Learned men with no axes to grind tell us, only, that there is a grave possibility. Since there is no final proof, we can go full steam ahead, ignoring the fact that there is no proof that these things will *not* happen.

'There is no final proof'. Wickenden was writing before the thalidomide babies. There is still no final proof but there is one horrifying example of the misuse of inadequately tested drugs.

What of the pellets of stilbestrol used for cattle and lambs?

'If ever a lamb-size pellet, or a steer-size pellet, finds its way to the family dining table, what is that biological bomb going to do?' asks Leonard Wickenden, recalling what the chicken-size pellets did to the mink.

In the rush to get ever increasing gain from ever decreasing amounts of feed, do we in fact gain? Wickenden thinks not:

> Even if competition reduces the price of meat, will the purchaser be receiving the same amount of nutriment for less money, or will he be getting more water in his beef? Does it make sense to claim that, by adding a trifling amount of chemical to cattle feed, 19 per cent more protein is produced from 11 per cent less feed? Has man *ever* been able to get something for nothing? We must insist on knowing where the extra weight comes from. If it is water, the consumer is the victim of what Dr Enders called an 'economic fraud'.

Even the butchers do not think the quality is there. Wickenden reports an article in *Farm Journal* in which they warn producers using stilbestrol that they may not get as much money for stilbestrol-fed cattle and add a Chicago meat packer's comment:

> It's not only stilbestrol that's responsible; it's the short-cut, cheaper fattening methods promoted by every agricultural college around. The beef we're seeing today doesn't measure up to the old corn-fed beef. It looks plump and good on the outside, but when you cut it open, the quality isn't there.

Finally I must quote a last fascinating report given by Wickenden on a meeting of the U.S. Department of Health, Education and Welfare, in Washington, D.C., on 24th January 1956, when:

> ... Granville F. Knight, M.D., W. Coda Martin, M.D., Rigoberto Iglesias, M.D. (of Chile), and William E. Smith, M.D., presented a paper entitled 'Possible Cancer Hazard Presented by Feeding Diethyl-stilbestrol to Cattle'. They stated that this powerful drug was known to induce cancer ... they stated that the drug was not destroyed by temperatures encountered during cooking. Yet it was known that the administration of the group of substances to which stilbestrol belongs has induced, in experimental

animals, polyps, fibroids, and cancers of the cervix, cancers of the breast, and serious pathological changes in sexual organs of male animals. Pellets removed from an animal one year after implantation were found to retain sufficient activity to induce a tumour upon being reimplanted in another animal. The effective dose, they stated, approaches the infinitesimal.
A continuing exposure to an extremely minute dose is found to be far more dangerous than intermittent injection of large doses. There are, on record, 17 cases of cancer of the breast in *men* who were given estrogens for treatment of prostate cancer.

But the U.S.D.A. says that it is quite all right to add them to the nation's meat.

Leonard Wickenden has, of course, been reporting the situation in America, but the same is true of this country. The use of hormones in poultry was made illegal in the United States in December 1959, but its use in cattle was still permitted. *The Times*, reporting on the banning of hormones by the New South Wales Cabinet, stated that the Australian Agricultural Council

> … was told that some countries, notably Italy, suspected that hormone residue in meat could cause malformation in children and abnormalities in adults, especially men. …
>
> A Ministry of Agriculture spokesman said yesterday that the possible hazards to consumers and workers had been carefully considered by the departments concerned and no evidence of harmful effects had been found. 'It has not been thought necessary to ban their use in this country,' the spokesman added. (13th November 1962)

So in this country we still carry on using hormones in beef calves, poultry and sheep.

The Ministry have now told us not to worry unduly about the use of either hormones or antibiotics.


## Chemical Insecticides

Chemical insecticides have been used with carefree abandon by farmer and housewife alike during the last twenty years, the housewife with her numerous aerosol sprays, and the farmer with giant machines and helicopters.

But the eradication of insect pests is not proving as facile a process as was at first taken for granted. The effect has been twofold. Weaker insects have indeed died, but the stronger ones have survived and built up immunity to the insecticide, breeding in turn progeny which have even greater immunity. The failure is highlighted by the fact that these chemicals fall on friend and foe alike, and we often lose friends and gain even stronger foes. Thus the song bird and the bees are killed while flies achieve a bouncing immunity.

This is the same vicious circle we have just seen in the relationship of drugs to disease. Increasingly powerful insecticides follow each other with great speed onto the market, and as a resistance develops in the insects, the race goes on. But we are steadily losing ground. The insects are becoming resistant to almost anything we put out, and there is not time owing to the speed with which we produce these chemicals for them to be tested for possible danger to humans. So we are faced with the situation of an indiscriminate use of some two hundred chemicals, all highly potent poisons.

The concentration of stock in a confined and enclosed area creates a humid and warm atmosphere ideal for the propagation of flies, red mite, lice and other pests. To combat these, powerful generators have been invented which eject chlorinated hydrocarbons and other insecticides as aerosols which 'advance as a gas disseminating over every exposed surface'. They are absorbed through the animal's skin or with its food and stored in its fat.

'There is no need to remove the birds when insecticide is being applied,' reads an account of a new appliance for insecticide fumigation. Accounts as to the effect on the birds seems to vary. 'Insecticides containing chlorine might be responsible for outbreaks of toxic fat disease which in the past two months have caused heavy losses in broiler chickens,' Dr Blount of B.O.C.M. is reported to have said (*Farming Express*, 20th July 1961), but *Farmer and Stockbreeder* reports research work being done in America to end fly infestation of poultry manure by feeding the birds insecticide (5th March 1963): 'U.S. workers have reported favourably … but research in Britain proved conflicting. No ill-effects on poultry health was reported but the fly-control measures failed.' Rachel Carson tells us that when fields of alfalfa are 'dusted with DDT, and meal is later prepared from the alfalfa and fed to hens; the hens lay eggs which contain DDT'. This, presumably, has not been considered in our efforts to get rid of flies from battery houses. Meanwhile plagues of flies from modern chicken houses are spreading all over the country and becoming a cause of anxiety to Medical Officers of Health. One such Officer explains that keeping chickens on wire is the cause of all the trouble because 'while flies have always used droppings as breeding ground, this did no harm in the old days because the birds ate the fly larvae which appeared' (*Poultry World*, 18th October 1962). He then went on to explain that with wire separating the chickens from their droppings this form of control was no longer possible.

> While the eggs are being laid up above 'in the racks', down below in the droppings pit the flies are breeding away at an alarming rate quite undisturbed….
>
> Tests were carried out in the Chichester area using a bacterial insecticide which causes a fatal disease in some flies, but the lesser house fly appeared to be immune.

Some planning committees are taking the only action left to them to protect the public from this upsurge of flies and are turning down applications for permission to keep battery houses anywhere near to dwelling houses.

There is equal difficulty with flies in other livestock sheds. A photograph of a Dutch veal house of the more modern type has a caption: 'Flies are everywhere. Look closely at this picture and you will see them lying dead in the gutter in their hundreds. Resistance to all known fly sprays is said to be developing in Holland' (*Farmer and Stockbreeder*, 13th September 1960).

Leonard Wickenden correlates the increase of hepatitis in man and hyperkeratosis in cattle with the increased use of insecticides, and quotes an article published in the *American Journal of Digestive Diseases*, November 1953, by Dr Morton Biskind, which suggests that the cause of many other diseases may be attributable to the same source:

> In animals, cattle have developed hyperkeratosis (or X disease) and the incidence of hoof and mouth disease has risen; sheep have 'blue tongue', 'scrapie' and 'over-eating disease'; hogs have vesicular exanthemata (a blistering of the skin); chickens have Newcastle disease and other ailments; dogs have developed 'hard pad' disease and the highly fatal 'hepatitis X'; and so on. With the obvious exception of hoof and mouth disease *not one of these conditions is mentioned* in the comprehensive U.S. Department of Agriculture Handbook, *Keeping Livestock Healthy*, published in 1942. This coincidence alone should have been sufficient to arouse suspicion that something new, that is common to man and his domestic animals, has been operating in their environment during the period that these changes have occurred.... (*Our Daily Poison*)

X disease in cattle, it appears, spread rapidly until by 1948 it had been reported in thirty-two states, and by 1949 was the cause of mortality in 80 per cent of calves under six months of age, 50-60 per cent of calves over six months of age, and 15 per cent in adult cattle. It was finally traced to poisoning by a chlorinated hydrocarbon. Wickenden then quotes another extract from the article by Dr Biskind giving confirmation to his thesis that chlorinated hydrocarbons have been a cause of increased hepatitis (inflammation of the liver) in man also, and this quotation tells of a hospital where an epidemic of hepatitis persisted among the resident staff for three years during which time chlordane had been used as a routine measure in the kitchen and food stores.

There is no generally accepted level at which DDT can be safely stored in the body and one of the most terrifying aspects of it is its power to magnify within the fat. Rachel Carson illustrates this very powerfully in her story of Clear Lake, California, much beloved of fishermen. It was decided to wipe out the gnats which marred the fishermen's pleasure and accordingly the lake was sprayed with DDD, a close relative of DDT but less toxic to fish. In 1949 the lake was sprayed with insecticide diluted to the rate of one part chemical to seventy million parts water. The spraying was successful until 1954 when the gnats returned. This time the spray was only diluted one part to fifty million of water. There was a third spraying in 1957 and large numbers of grebe, which normally thrived round the lake, died off. Then a curious thing was discovered. The plankton in the lake picked up the DDD and handed it on from generation to generation, long after the water was clear. This was eaten by the fish, which were found to have 300 parts per million stored in their fat. These fish, in turn, were eaten by the grebe, which had a concentration of 1,600 parts per million in their fat – and died.

It is as well to remember that we are at the end of a similar biological chain.

We have had similar experiences in this country. The Animal Health Trust tell of a case where poultry fed on seeds which had been dressed with chemical ready for planting. The poultry were so seriously affected that 70 per cent of them died. Fox cubs eating the birds then also died. Perhaps it is fortunate that we don't eat fox cubs.

It is going to take many generations even to begin to form a picture of what tolerance is permissible in man before noticeable breakdowns occur. Meanwhile the effects of acute doses in man and in animals have indicated that 'one would expect hyper-excitability of the nervous system and damage to the brain, liver, kidneys, suprarenals, thyroid, endocrine system in general, and loss of appetite. Possibly the bone marrow could also be damaged' (Franklin Bicknell).

Dr Bicknell points out also the risk attached to the possibility of the chemical being passed from the mother to the unborn foetus in her womb, and through her milk to the infant at her breast, both possibly causing damage as yet unexplored.

It is in times of illness that the stored insecticide in the fat can cause most damage. When the illness is prolonged and serious most of the fat may be rapidly used up and the DDT stored up in it is suddenly released into the body, far more toxic to it in its weakened state.

Rachel Carson has linked the dangers of insecticides with that of synthetic hormones, just one illustration of the possible dangers arising from the interplay of the many chemicals which we store in our bodies. She shows us that the body's built-in protection against an excess of natural oestrogens is in the liver. But if the liver is damaged, or the body's supply of B vitamins reduced, this protection is lost and the oestrogens build up to abnormally high levels.

> In brief, the argument for the indirect role of pesticides in cancer is based on their proven ability to damage the liver and to reduce the supply of B vitamins, thus leading to an increase in the 'endogenous' oestrogens, or those produced by the body itself. Added to these are the wide variety of synthetic oestrogens to which we are increasingly exposed – those in cosmetics, drugs, foods and occupational exposures. The combined effect is a matter that warrants the most serious concern.

It is not that we do not recognise *any* danger. It is sufficiently recognised by the agricultural world for warnings to be given as to how chemicals should be handled. A Ministry leaflet is ironically called 'The Safe

Use of Poisonous Chemicals on the Farm', and advises, amongst other things, that the workers wear masks and overalls whilst spraying, that they never allow the chemical to touch their hands or food, and that they wash thoroughly when they have finished. Details are given of first aid to be applied, until medical help can be obtained, if the worker even then gets poisoned; if he feels ill, has weakened breathing, or gets convulsions. To this extent the sprays are acknowledged as potently dangerous, but at this stage the acknowledgement of their dangers in this country comes to an abrupt halt.

In New Zealand the Government was made to face up to the subject when a shipment of their meat was measured by the United States health authorities and a chemical residue above the U.S. safety level found. At once the government took action and banned the use of 130 brands of pesticide on livestock.

> The banned products were those containing aldrin, dieldrin, BHC, lindane, DDT and methoxychlor, and included a large number manufactured in Britain and available to British farmers.
>
> The New Zealand government claimed that these chemicals tend to leave residues in the products of treated livestock and the *suitable and effective alternatives are available*.

The report in *Farmer's Weekly*, 9th February 1962, goes on to say that Australia is anxious and deciding what precautionary measures she should take. It finishes by giving us but very cold comfort:

> A leading British manufacturer has formulated alternative insecticides *for export to New Zealand*. A spokesman for the firm said that if similar restrictions come into force here, effective alternatives would probably be more expensive than existing chemicals. (The italics are mine.)

The attitude of our own Ministry of Agriculture is, as always, reassuring. Dr Sanders, speaking in a B.B.C. television programme on 'A Suspicion of Poison' at the beginning of 1963, assured us that there was strict control in this country. He had found no case of dangerous level of pesticide left in food, in fact residues were only a hundredth of what is dangerous. But are not these chemicals stored and magnified in the consumer's fat, together with residues of drugs, hormones and other chemicals used in processing and preserving food?

Lord Hailsham, Minister for Science, was reported in the *Financial Times* (21st March 1963) as deploring alarmist attitudes against the use of chemicals in agriculture and preservation of food and as suggesting that 'we could not have the benefits of a scientific technological society without running some of the risks'.

Why not? you may feel inclined to ask, for does not New Zealand show us that the only risk we need run is a monetary one, and we might make a capital gain out of that in time.

## Taste and Quality

At the 1962 World Poultry Congress at Sydney, the World Poultry Science Association President, Dr H. H. Alp, shook delegates by declaring that research in the poultry world was not proceeding along broad enough lines:

> As an example of the manner in which current research was failing he said that most of the nutritional research now being conducted was centred around the problems of getting more meat or more eggs from each pound of feed. *'What is the use of this,'* he asked, *'if the additional weight of meat was tasteless and the eggs so inferior in quality that less of them are eaten?'* ... *'The world,'* declared Dr Alp, *'cannot buy research and the world of science, with the World's Poultry Science Association, must maintain intellectual honesty and be free from bias.'* (*Poultry World*, 16th August 1962)

When the word quality is mentioned, it is usually intended to denote uniformity of size and texture, freedom from blemishes and other superficial characteristics, and its correlation with nutritional quality is very rare. It is natural to suppose, however, that there must be some correlation between taste and quality in the nutritional sense or how else could we have survived all these thousands of years when our only guide to food has been its taste and palatability? Dr Milton puts the case better than I can:

> Palatability is not a quality which is easily defined. We know it when we come across it. The taste-buds are satisfied at the back of the tongue. We find that the food is worth chewing, and we feel that we would like more. This *must* be a nutritive quality because there is a sense of fulfilment of desire brought about by eating such food; and if we are happy in eating our food, I am quite sure that it is much more likely to produce good health. On the other hand, and by contrast, if we eat something that looks good but doesn't taste good when it is in the mouth, then a feeling of disappointment arises that runs through the whole organism; it affects the glandular system, the digestion becomes upset, and a chain reaction is produced which can be quite harmful to the body.... (*Mother Earth* October 1962)

A French veterinary surgeon, Michel Perrin, says rather more succinctly that according to Pavlow the taste of a food is proportionate to its digestibility and has a considerable influence on its nutritional value. An empirical appreciation of this, Perrin says, amounts to the same result: of meat which is more tender, but has less flavour or nutritional and digestive value, 'it is said that it satisfies hunger less and gives less energy'.

In the historical sense, and until quite recently, I should have thought that this axiom of Pavlow's was self-evident, but the recent advances of the chemist would seem to raise a doubt whether taste can any longer be relied upon as a touchstone of quality, because it can now be synthesised at will.

Broilers have only one fault; they really taste of nothing (says Robin Clapham in *Farmer and Stockbreeder*, 23rd May 1961). I have been saying this since the first ten-weeks chick came out of the first-ever intensive plant… the problem has even now come to the ears of broiler growers themselves … no less a person than Dr W. P. Blount has been quoted as saying that the big groan up and down the country is the fact that broilers are tasteless.

He goes on to say that the industry will have to give the chickens some taste:

If people say they will only eat rhubarb-flavoured chicken then there is no point in kicking against it – we've got to sell them what they want…. It is not inconceivable that the American threat to 'Bury Us in Chicken' will be supported by the introduction of novel flavours in wide variety….

The *Financial Times* (11th October 1962) reports what must be the acme of achievement in this field – a chicken flavour for chickens.

… The industry which has conquered the pallid yolk resulting from intensive laying (by the introduction of certain dyes into feed) is now experimenting with a 'chicken taste' extract to meet criticisms that intensive rearing leads to tasteless meat.

To be followed no doubt by bacon, beef and lamb tastes.

Professor Yudkin, of Queen Elizabeth College, London, speaking at the Royal Institute of Public Health and Hygiene Conference in 1962 on 'Hazards of Life' takes us further into this chemical jungle:

But the food industry has other capabilities. It can make extracts of foods, and mixtures, which, by alterations of texture and additions of colour and flavour, we find highly palatable. And I am now going to suggest that this ability of the food manufacturer to make for us new foods, which satisfy only our demand for palatability, is a dietary hazard of the greatest importance.

… By eating foods for palatability, we eat foods which give us our nutrients. But this is true only so long as we have not separated palatability from nutritional desirability. Now, thanks to modern technology, the food manufacturer can do just this. We demand of him that he produce foods which look nice, taste nice and have a pleasant texture. Since we demand nothing more, and since he can provide us quite readily with these qualities of palatability quite apart from qualities of nutritional value, this is what we frequently get.

… We may soon be eating pies, hamburgers and sausages, with *every quality of the meat they should contain, except the nutritional value.* (The italics are mine.)

In view of this it is clear that there is much room for research on the nutritional effects of those forced foods. In fact surprisingly little research has been done on the nutritional value of the forced foods which form the subject of this book, and the main reason for this is that money has not been forthcoming to enable independent and unattached scientists to undertake the necessary longterm studies. Let me not be misunderstood. There appears to be no lack of funds for research into food problems, but almost all this research effort is tied up in one way or another to the sponsoring industries and very little is available to provide unbiased answers on behalf of the consumer.

From this pitifully narrow field I have selected three small items of research carried out on eggs by scientists of independent mind and, I believe, unbiased judgement.

The first is a survey carried out by Dr Milton, a leading analyst in the food trade. He made the analyses every month for a year in 1959 and his findings were:

*Battery eggs*

| | |
|---|---|
| Vitamin A | 4,200 international units per 100 g. |
| Beta carotene | 310 international units per 100 g. |
| | ——— |
| Total | 4,510 international units per 100 g. |

*Free range eggs*

| | |
|---|---|
| Vitamin A | 7,200 international units per 100 g. |
| Beta carotene | 1,630 international units per 100 g. |
| | ——— |
| Total | 8,830 international units per 100 g. |

Beta carotene is the substance converted by the body into Vitamin A. Vitamin A is the anti-infection vitamin, the one that helps to give us protection against influenza etc.

Laurence Easterbrook, from whom this analysis came, comments:

> It is a far better guide to nutritive quality than protein, which is a purely chemical analysis telling us practically nothing.

And in August 1961 he confirmed that these figures were still unchallenged and added, 'Dr Milton has now gone a bit further and it would appear that whatever you put in the birds' food does not compensate in nutritional value for keeping them under such (battery) conditions.'

Dr Frank Wokes, of the Vegetarian Research Centre, Watford, published a paper with F. W. Norris from the Department of Biochemistry, University of Birmingham, on B vitamins in foods, in which they gave results of some research which they too had done with free range and battery eggs, showing the importance of the egg as a part of the human diet:

> … In human diets significant amounts of $B_{12}$ may be provided by hens' eggs. Recent authoritative results (McCance and Widdowson, 1960) indicate that about half of the suggested human daily adult requirement of 1 $\mu$g. can be obtained from one average fresh egg, provided that the hens have been

fed a normal English commercial breeders ration containing animal protein and about 1·2 $\mu g$. B$_{12}$ per 100 g. However, many hens (e.g. in batteries) receive much less B$_{12}$ in their rations and their eggs may contain only a quarter or a fifth of the daily human requirement. On the other hand, when the hens are obtaining ample supplies of B$_{12}$ from animal matter or suitable micro-organisms in the soil or manure, as in free range or deep litter production, each of their eggs may provide on the average rather more than the daily human requirement…. This value, however, decreased during storage under normal conditions as predicted by McCance and Widdowson (1960).

*Vitamin B$_{12}$ content of eggs ($\mu g./egg.$)*

|  |  |  | *mean contents* |  |
|---|---|---|---|---|
| *Free Range* | July | 1·24 | ± | 0·15 |
|  | August | 0·83 | ± | 0·03 |
|  | September | 0·70 | ± | 0·16 |
|  | October | 1·05 | ± | 0·03 |
|  | November | 0·48 | ± | 0·04 |
| Mean |  | 0·77 |  |  |
| *Deep Litter* | July | 1·28 | ± | 0·18 |
| *Battery* | July | 0·43 | ± | 0·02 |
|  | January | 0·34 | ± | 0·04 |
| Mean |  | 0·39 |  |  |
| *Shop* |  |  |  |  |
| 'farm eggs' | November | 0·53 | ± | 0·07 |
| ordinary | December | 0·27 | ± | 0·01 |

Means of 12 eggs, all known to be new laid except the shop eggs.

From these results you will see that the battery egg has appreciably less of the very important vitamin B$_{12}$ than the eggs from hens reared in more normal conditions. They are small experiments and on that account cannot be regarded as conclusive in themselves, but they do point to the need for much more massive work on the subject and for the funds to enable such work to be undertaken.

Lastly I would like to quote research done by the Oxford nutritionist, Dr Hugh Sinclair, who informed me:

> I did a little work a few years ago in my own private laboratory, which indicated to me that battery eggs when hatched produced arteriosclerosis in the day-old chicks, whereas eggs from free-ranging hens did not.

He enlarged on this experiment in *The Lancet*, 28th January 1961:

> My preference, on present evidence, for eggs from farmyard rather than battery hens was based on a simple experiment I did in the summer of 1956. I allowed the eggs of each type to be fertilised, hatched the chicks, and found material staining red with Sudan IV in the aortas of the day-old chicks from eggs of battery hens and not those of free-ranging hens. Some of the chicks

of the former type were then reared indoors on a commercial meal, and in these when killed the fatty deposits were greater whereas there were none in the chicks of the same origin reared free-ranging. I should like to follow up this work … because eggs are a very important item in our dietaries and have come under suspicion, perhaps wrongly, because of their efficiency in raising serum-cholesterol.

The steady rise of arteriosclerosis in the country must make us examine any methods of reducing it and take seriously any research which throws light on the problem.

Dr Bicknell explains the importance of E.F.A. (essential fatty acids) in our diet:

Cholesterol is a substance closely akin to Vitamin D, to the sex hormones, to the hormones of the suprarenal cortex and to the cancer-causing hydrocarbons … a diet rich in solid animal fats or in hydrogenated oils like margarine is one of the main causes of a high level of cholesterol in the blood, while the liquid vegetable oils containing essential fatty acids or E.F.A. reverse the high level caused by the solid fats…. hydrogenating oil not only destroys its E.F.A. but converts it into an anti-E.F.A., thus increasing the requirements of the body which already are ill met by our diet. Ill met because the E.F.A. which should be provided by pigmeat, eggs and milk may be absent because the concentrates fed to pigs, cows and hens may lack E.F.A. and so cannot be handed on to us.

The Ministry themselves, by inference, admit the inferiority of intensively produced eggs. In *Incubation and Hatchery Practice* they point out that:

… the survival and successful development of the embryo, isolated as it is from any day-to-day replenishment of its food stores, must depend on the egg containing the right amount of every nutritional necessity except gaseous oxygen….

The fact that commercial laying hens will produce well on a variety of laying diets should not be allowed to give the impression that the same diets are adequate for breeding stock. Emphatically this is not so, and the use of such diets will soon lead to troubles. Furthermore, the fact that breeding stock are laying eggs is no guarantee that even if these eggs are fertile they will hatch, or, if they do hatch, that the chicks will be able to develop properly. The usual limiting factors are deficiencies of vitamins or minerals in the egg, and it must be realised that the degree of deficiency that is enough to render hatchability a matter of speculation need not have any ill effect on the health or the productive performance of the hen.

Dr Milton defines quality in food as 'a factor which will allow a condition of positive health to be maintained in the organism' and this may be the only means we have left of defining quality, after the natural taste has been taken out of food and resubstituted by synthetics, and after its palatability and texture have been played upon by drugs and hormones.

# Cruelty and Legislation

Any discussion of cruelty to animals can be obscured by two equally unsound bases of thought. First, there is a very large body of animal lovers who anthropomorphise the animal and tend, not only to regard animals and man as equals in every way, but to put animals before man in their concern and regard. These people do more harm than good to the very cause they seek to help. Then there are those who pride themselves on being sensible and unsentimental in their approach, and who consequently allow to animals little sensitivity and less intelligence, designating them as being in existence solely for man's use. These brush off any sense of personal responsibility if this usage entails suffering to the animal, unless it be of such harshness that even they may safely condemn.

It is because there are so many degrees of cruelty, and so many shades of acceptance of cruelty, that laws have to be passed and rigidly enforced, even in our 'civilised' times.

That animals are sensitive to pain and discomfort is obvious from the fact that in the higher animals the sensory and nervous systems are similar to those of man. An animal of higher intelligence would have a greater span of memory and sense of anticipation and therefore suffer more acutely than less highly developed species, but the initial pain felt would be experienced equally by both. Dr John Baker, Lecturer in Zoology in the University of Oxford, elaborates this point:

> It is probable that there is some degree of correlation between intelligence and capacity to suffer. It is thought that intelligent animals feel pleasure and pain, recognise and remember the concomitants of these

© J. Harrison and J. Wilson 2013. *Animal Machines* (Ruth Harrison)

feelings, and modify their behaviour accordingly. It is true that an intelligent and a foolish animal might feel pain equally, but modify their behaviour as a result to very different degrees – the foolish one perhaps not at all; but though the immediate perception of a painful stimulus might be equal in the two cases, yet the more intelligent animal would be the more prone to suffer from memory and hence from apprehension. In general, it is reasonable to assume that the larger and more complex the cerebral hemispheres are, the greater is the probability that the animal is capable of suffering acutely. (*The Scientific Basis of Kindness to Animals*)

Can an animal's reluctance to return to a situation which has been the source of discomfort or pain be put down purely to memory? In the maintenance of the pecking order, for example, a bird must remember which other birds are above and which below it in the social order, and although the original struggle may have been painless, the social order, once established, is not quickly broken by any bird. But this would also seem to denote a sense of anticipation being experienced by the bird. It does not venture to cross the path of its conqueror because it is afraid of, being pecked and hurt. It is true that the hen kept in a battery cage has not the same chance as the free range bird to develop even a degree of intelligence, but I would not deprive this, the lowest form of stock which comes within the scope of my book, of the powers of pleasure and fright which give variety to its life and make the distinction between an organism which has consciousness and a plant or machine.

There is a vague impression throughout people in this country that animals are protected against all forms of cruelty by the law. Let us examine that law however. The most important legislation which is supposed to protect animals from cruelty is The Protection of Animals Act (1911). This Act is, of course, hopelessly outdated. Its drafters could not possibly have envisaged the type of insidious cruelties which are perpetuated in modern animal husbandry. Let us now examine the clauses of The Protection of Animals Act 1911 so that it may be seen how difficult it is to protect against modern forms of cruelty within the meaning of the Act:

**1.** –(1) If any person –

    (*a*) shall cruelly beat, kick, ill-treat, over-ride, over-drive, over-load, torture, infuriate, or terrify any animal, or shall cause or procure, or, being the owner, permit any animal to be so used, or shall, by wantonly or unreasonably doing or omitting to do any act, or causing or procuring the commission or omission of any act, cause any unnecessary suffering, or, being the owner, permit any unnecessary suffering to be so caused to any animal; or

    (*b*) shall convey or carry, or cause or procure, or being the owner, permit to be conveyed or carried, any animal in such a manner or position as to cause that animal any unneccessary suffering; or (*c*) shall cause, procure, or assist at the fighting or baiting of any animal…

(*d*) shall wilfully, without any reasonable cause or excuse, administer, or cause or procure, or being the owner permit, such administration of, any poisonous or injurious drug or substance to any animal, or shall wilfully, without any reasonable cause or excuse, cause any such substance to be taken by any animal; or

(*e*) shall subject, or cause or procure, or being the owner permit, to be subjected, any animal to any operation which is performed without due care and humanity;

such person shall be guilty of an offence of cruelty within the meaning of this Act, and shall be liable upon summary conviction to …

–(2) For the purpose of this section, an owner shall be deemed to have permitted cruelty within the meaning of this Act if he shall have failed to exercise reasonable care and supervision in respect of the protection of the animal therefrom….

–(3) Nothing in this section … shall apply –

(*a*) to the commission or omission of any act in the course of the destruction, or the preparation for destruction, of any animal as food for mankind, unless such destruction or such preparation was accompanied by the infliction of unnecessary suffering; or

(*b*) to the coursing or hunting of any captive animal….

Apart from 1(*c*) and 3(*b*), these are the clauses of concern to us in the discussion of their relevance to intensive rearing methods and to their adequacy to present day circumstances.

It is worth noticing that cruelty is never defined as an absolute quality, there is plenty of scope for interpretation in the Act. What strikes one forcibly is also the reluctance to define cruelty in relation to animals reared and killed for meat. In fact if one person is unkind to an animal it is considered to be cruelty, but where a lot of people are unkind to a lot of animals, especially in the name of commerce, the cruelty is condoned and, once large sums of money are at stake, will be defended to the last by otherwise intelligent people.

The looseness of definition in the Statutory Law would not be of such importance if cases had been brought under Common Law to act as precedents. But this has never been done in relation to animals. I was told by a senior official of the Royal Society for the Prevention of Cruelty to Animals that the Society never takes up a case it is not 100 per cent sure of winning, and this presumably is why it has never taken action under Common Law. At the same time I would have thought that these Common Law actions, even if not 100 per cent successful, were more effective in pin-pointing deficiencies and discrepancies in the Statutory Law than any amount of discussion in the press. The fact that no action has been taken against any farmer, even those carrying intensivism to extremes, has allowed successive Ministers of Agriculture safely and smugly to give a set, or 'bugs', answer systematically to every critic:

> Whether these conditions would amount to cruelty would be a matter for the Courts to decide under the provisions of the Protection of Animals Act 1911….

The Ministry seem to presuppose in such statements that the development of this sort of rearing since the Act was passed does not make necessary any revision of the Act. They have also asserted that the animals thrive and there is therefore no reason to suppose that they suffer any discomfort, that all farming is an interference to some extent with nature, and that they have not heard that extreme conditions prevail in this country anyhow. With which contradiction in terms we may safely leave them.

The extent to which farming has now carried its 'interference with nature' has reached far beyond depriving the animal of its birthright of freedom, sunlight and green fields. It has now reached the point of frustrating practically every natural instinct in animals – except the instinct for survival. Even this only exists precariously in some of the intensively reared animals. Mr Stephen Williams of Boots' farms, told a meeting in discussing general problems created by intensification that his 'observations suggest that something happens to the animals'. He thought 'they had less will to live and resist disease….' and *Farmer and Stockbreeder* reported him as saying further that 'keeping livestock in larger groups meant that the more timid animals became the submerged individuals, and the bigger the group, the greater the tendency for some of them to commit suicide' (*Farmer and Stockbreeder*, 30th January 1962).

The first instinct the farmer frustrates in all animals except pigs, is that of the new born animal turning to its mother for protection and comfort and, in some cases, for food. The chick comes out of the incubator and never sees a hen, the calf which is to be fattened for veal or beef is taken from the cow at birth, or very soon after, and even the piglet is weaned far earlier now than it used to be. The factors controlling this are mainly economic.

The next instinct we thwart is that of the animal to select its own food. Even outdoors within the confines of the field boundaries the animal cannot find a complete diet – especially now that we are eliminating with herbicides all the richness of the hedgerows, but there is obvious pleasure for the animal in its foraging and under good management this is satisfactorily supplemented with other food. The American naturalist, Roy Bedichek, points out that in the battery hen it is not only the scratching instinct which is frustrated, but also the 'nervous and muscular mechanism which responds to it' (*Adventures with a Naturalist*).

That the eternal pellets of the manufactured diet are not always either satisfactory or satisfying for the animal is illustrated by the fact that baby beef calves 'gnaw at the woodwork of their pens, so much do they crave roughage' (*Farmer and Stockbreeder*, 10th April 1962), and the veal calves lick and suck at everything within reach including the urine from their slats. By immobilising animals and placing before them a monotonous supply of the same food, we have taken away an interest in life and all these intensively reared animals suffer acutely from boredom. Animals are essentially curious about what is going on and enjoy watching the world go by almost as much as we do. Bedichek points out that hens in battery houses, 'have stampedes. With no apparent cause a wave of hysteria sweeps over the whole battery: wild, unnatural chirps, jumbled screams, and a fluttering as if every feather on every chicken had become possessed and frantic.' He then offers a possible explanation for this:

> The nervous mechanism of the chick – and of every member of the family of birds to which he belongs, for that matter – is geared to successive frights followed by periods of security…. It is the perversion of this instinct, I think, which accounts for the sudden and unmotivated panics…. As is the way with biologically determined habits denied expression and use, they exhibit themselves sooner or later in abnormal behaviour.

Boredom is a fault of intensivism that, by the very nature of the overcrowded units, producers find most difficult to eliminate. It can, and does, lead to 'vice'. Feather-pecking and cannibalism amongst birds, fighting and tail-biting amongst pigs, are the every day worry of the producer for damaged carcases can mean serious loss of profits.

No domesticated animal will choose naturally to lie in or near its own dung. It is wonderful to see a day-old piglet rushing anxiously to the dunging passage. Yet this escape from its own dung is something we have denied our intensively kept animals. Indeed we encourage their existence over dung as a virtue – economic of course, it saves fuel by keeping the animals warm. It is even considered an advantage to feed pigs kept on solid floors on the floor so that they can lick it clean and thereby save the space of food troughs and a chore to the farmer.

The natural sleep cycle is most rudely ignored. Experimentation is always on hand with lighting patterns for battery hens. The whole function of the hen is to lay eggs, and to this end they are kept in whatever amount of light is deemed suitable at the moment. At one farm a period of twenty-three hours lighting a day has been tested. The broiler chicken spends two-thirds of its life in darkness, the veal calf often spends its life in dim light if not in actual darkness, the new baby beef units encourage darkness, and pigs are often kept in dark houses simply because it is cheaper to build and run piggeries without lights.

But the Minister states that any cruelty is covered by the Protection of Animals Act, 1911.

Lastly, animals, and especially cloven hoofed animals, cleave to solid ground. They are uncomfortable and ill at ease on slats, yet for reasons of economy intensive reared animals are now nearly all kept on slats.

In 1961 the use of slats was heatedly debated all over the country and the Royal Show made a special feature of cloven hoofed animals on slats so that farmers could study them and decide for themselves the wisdom of using them. Controversy raged in the agricultural press. The Editor of *Farmer and Stockbreeder* summed up the argument and then put a suggested solution, one that typifies the contemporary attitude to stock:

> At the moment controversy rages over one of the highlights of the Royal demonstration site – slatted floors for stock. Save straw, save labour, stock are happy on them, says one side; bad for legs, draughty, undesirable, says the other....
>
> The open mind weighs up these arguments ... the commonsense approach at this stage in our knowledge is that *for expendable stock* the slatted floor seems to have more merit than disadvantage. *The animal will usually be slaughtered before serious deformity sets in.*
>
> On the other hand, breeding stock, with a long working life before it, *must grow and keep good legs*; and risk of damage here or from treading in the closer stocking which slats permit would seem to outweigh all the advantages. (11th July 1961. The italics are mine.)

You will notice from the photograph of veal calves in Figure 17 that these 'expendable stock' have fatty deposits round their leg joints due to lying on slats.

But the Minister states that any cruelty is covered by the Protection of Animals Act, 1911.

The *Farmer and Stockbreeder* Vet. made a study of two sets of cows, fed and looked after identically for two years, except that one lot was housed on slats and the other in a normal cowshed. Both sets were turned out into adjacent yards for his inspection:

> The comparison was almost unbelievable. On an average I'd say that the cowshed cows must have weighed $1\frac{1}{2}$ to 2 cwt. per head more than their slatted mates; and whereas the traditionally housed cows were bright-eyed, young and vigorous looking, the 'slats' were apathetic, tired and aged looking. Most of them were stiff: many had swollen fetlocks and hocks, and several had dropped flexor tendons. All of them moved about the yard apprehensively as though on a bed of hot bricks.

He reported that milk yields of the slatted floor cows were appreciably lower than the others. Similar difficulties had been observed with pigs housed on slats.

> Quite apart from anxiety neurosis (the Vet. continued), there is a physiological reason why cloven-hoofed animals should not be allowed on slatted floors. The entire mechanism of the tendons of bovine, ovine and

swine legs is geared to deal with a claw divided into two halves, and for this mechanism to function properly it is essential that the two halves be placed confidently on a reasonably uniform surface so that the weight of the animal is evenly distributed.

*Each time the double digit comes down on the slats there is inevitably uneven weight distribution, and undue strain is placed on all tendons and joints. Naturally, therefore, abnormalities result. (Farmer and Stockbreeder,,* 20th June 1961)

From agriculturists all over the country reports were published of the reluctance of cattle to go on slats where they could avoid it; one report said the cattle would rather lie in silage slurry than go on the slats. *Agriculture,* December 1961, had a report on work done at the National Institute for Research in Dairying on the usage of slats for dairy cows. They found there that cows did not lie down on the slatted floor for the first forty-eight hours and far less even after that time than did cows comfortably housed on straw. Unlike the *Farmer and Stockbreeder* Vet., however, they observed that:

> Although animals have sustained injuries to legs, hoofs and teats, there have been very few reports of reduced milk yields when cows have been introduced to slats, and there have been no reports of lower liveweight gains to store and fattening cattle.

The commercial advantages to producers won the day, and now not only is a large proportion of indoor housing on slats, but consideration is being taken a stage further to the use of a metal grid which, it is suggested, would make cleaning easier still. A farmer who has already tried these floors for his calves found that 'the calves lie down readily enough but appear to find the floor rather hard on the knees as they get up … after two days on this metal floor all the calves decided it was far more comfortable to get up "horse fashion" – front legs first' (*Farmer's Weekly,* 14th December 1962).

Because the commercial advantages of slatting are too great to be lightly dismissed, a successful experiment at the University College of Wales was quoted in *Agriculture,* August 1962, as being a useful compromise. The feeding area for cows was slatted to save on straw and cleaning out, and a bedding area of straw retained:

> All cows housed in this system were given access to their hay simultaneously. They lined up with very little fuss. The absence of bullying during feeding has been remarkable – possibly because all cows have much less confidence on slats than on other types of flooring. It is interesting to note that once they had finished their hay ration the cows returned at once to the bedded area. At no time during the total period of nine months in which the system has been in use has any animal been seen lying on slats. Only very occasionally indeed has an odd one been found at night standing on the platform, and there can be no doubt at all that, given a free choice, cows will not stay on slats.

The *Farmer and Stockbreeder* Vet. points out, finally, that the widely used cattle grid was devised expressly for its unsuitability for cloven-hoofed animals:

> When we want to stop sheep, pigs and cattle from getting out of a field and we don't want the bugbear of constantly opening and shutting gates, what do we do? We dig a pit, brick or concrete the sides and cover the hole over with a 'cattle grid' or, if you like, a 'slatted floor'. That done, we don't worry any more. Why? Because given the choice, no cloven-hoofed animal will ever willingly step on to a grid or a slatted floor, and never at any time will stock stray over a grid. (20th June 1961)

The proportion of void to solid in cattle grids is devised to render them untenable to the animals. The proportions in slatted floors are devised to render them just sufficiently tenable. A grid with close enough mesh to simulate a solid floor would be insufficiently effective in disposing of dung. Herein lies the rub. With the success geneticists are having in breeding featherless, neckless, almost legless birds, it should not be beyond their powers to devise a cow with the foot of a camel and so solve the problem once and for all.

The Minister obviously does not recognise this form of cruelty to cloven-hoofed animals for in June 1963 concrete slats of approved design were made eligible for a grant for all animals except dairy cattle.

The greatest condemnation of intensive methods of animal rearing is that *the animals do not live before they die*, they only exist. There is no longer any warmth in this business approach to livestock rearing, many producers state quite frankly that they hate their stock. The animal is the immediate loser in this state of affairs because the domesticated animal is very dependent on those who look after it. It is often said in the agricultural world that drugs have taken the place of stockmanship, but it is difficult to blame the stockman entirely for this. When he has vast numbers of animals to look after he cannot be expected to have the same feeling and instinct for their needs as he did with relatively few.

Nor can he be expected to see when something is wrong when the units are in dim light or darkness. It is asking too much of the stockman and it is placing a great strain of endurance on the animal.

> Isolation and treatment of individual sick birds is not practicable and mass medication by the use of feed additives is the only suitable method of controlling disease outbreaks,

we are told by a drug firm. But an article in *The Countryman*, Autumn 1960, points out a further hazard:

> … There will always be some animals or birds which do not fit into the system and, in 'factory' conditions, when comparatively large numbers are involved and the margin of profit is small, they are liable to receive scant consideration. Fortunately this is less likely to occur with a calf, which will

be worth a great deal more than a bird; but broiler sheds certainly exist where ailing birds are left to die unnoticed.

Dr Blount of B.O.C.M. admits in his book *Hen Batteries* that:

> Some birds, fortunately only about one in a thousand, do *not* take to cages – they refuse to eat or drink … they lose weight, and if left alone would eventually die, emaciated. (Remember, however, that these 'sulking' birds are readily detectable and can be culled quickly….) Post mortem examination reveals no abnormality. *Such birds are unhappy*, and to retain them in cages is cruel….

As a result, twenty-eight thousand birds a year have to be killed simply because they are unhappy when caged.

But the Minister states that any cruelty is covered by the Protection of Animals Act, 1911.

For hundreds of years the domesticated farm animal has been moulded to man's purpose to produce for him just what he wants in ever faster and cheaper ways. It is surely unlikely that even under the adverse conditions it has to endure today the inherent mechanism will immediately crumble. This will take time. Meanwhile the constant defence put forward by the producer and his supporters is that the animal thrives and would not do this if it were suffering, and that indeed he would not make any profit unless the animal did thrive. But how far is this true?

Now an accepted connotation of thriving is that of flourishing, a factor of health, and on this basis it is impossible to think of these animals thriving. In the case of the veal calf, for example, the producer makes his profit on the calf's anaemia, on its *lack of health*. The Ministry of Agriculture admits that anaemia is accepted as being inevitable in the production of 'white' flesh. Some broiler beef calves are coming out of their year's confinement blind, with damaged livers, and other complaints. Even the blind animals however, are graded as 'A' quality in the slaughterhouse.

But in health, as in law, we find a discrepancy in definition regarding animals as against man. The Oxford Dictionary defines thriving as 'prosper, flourish, grow rich'; but of animals 'grow vigorously', and this a young animal will do under almost any conditions, we are told by Dr Stephen Bartlett, recent head of the Dairy Husbandry Department of the National Institute for Research in Dairying:

> Growth and development of living animals is a very complicated process and simple explanations of it are liable to be distortions. But it is true to say that one outstanding feature of all new-born animals is a *tendency to grow at all costs*. Semi-starvation may check growth but will not stop it…. (The italics are mine. *Farmer and Stockbreeder*, 7th March 1961)

I think, however, that it would be fair to couple this natural tendency to 'grow at all costs' with these animals' inability to do other than put on weight, for that is their sole purpose in life, and their management is arranged entirely to that end. All their energy must go into converting food into flesh and it is hardly surprising that they put on weight. But this has nothing to do with contentment or health. Would contented animals have such a sweeping mortality rate that they must be kept alive with continual antibiotics? Why, if the beef calves are happy in their fifteen square feet of space, must they be given tranquillisers in their food? And isn't their putting on weight rapidly partly due to the questionable practice of feeding hormones to them?

Laurence Easterbrook, contributor to the *Daily Mail*, wrote in July 1961:

> Rightly or wrongly, I hate the idea of keeping hens in those wire chests of drawers they call batteries. I think it is cruel and also that it produces an inferior article of food.
>
> It's no good telling me that they wouldn't thrive if they were unhappy. The time when I really put on weight was when I sat in trenches for a year, dirty, wet, frightened, and miserable.

It is obvious that the producer measures the animal's comfort and happiness solely in relation to his profits, and while the animal is willing to endure and exist as part of this machine, he will defend to the last his right to use these methods.

I would like to put before you some of the other ways in which the law, as it stands, fails these animals, and where also, I feel that test cases might have been taken and won.

Clause 8 of the Protection of Birds Act, 1954, states:

> (1) If any person keeps or confines any bird whatsoever in any cage or other receptacle which is not sufficient in height, length or breadth to permit the bird to stretch its wings freely, he shall be guilty of an offence against this Act and be liable to a special penalty:
> *Provided that this subsection shall not apply to poultry....* (My italics)

What are we to make of a law which permits all caged birds whatsoever to flap their wings *except poultry*? A law such as this brings derision on itself and on its enactors alike for its manifest casuistry. Well may we be called in contempt a 'nation of shopkeepers'.

This proviso has allowed the battery industry under its legal aegis progressively to reduce the living space per bird to a point where not only is it unable to stretch its wings but it cannot even stretch its neck.

Makers of battery cages made a revolutionary move in 1962 under the slogan 'Why Waste Space?' and dropped the height of their cages from 7 ft. 11 in. for a *four tier unit* down to 6 ft. $5\frac{1}{2}$ in. This made the cage itself (remembering the spaces between the cages for the droppings belts), 15 in. high at the front and only 12 in. high at the back. To allow eggs to roll away the bird is condemned to live out its existence on a 1 in 5 slope, which accounts for the difference between the height at the front and at the back. Try to imagine the floor of your living room with a slope of 1 in 5 and the muscular strain you would be subjected to in trying constantly to maintain your balance. 15 in. high by $15\frac{3}{4}$ in. wide by 16 in. deep, and into this space go three small hens or even four. Take a rule and measure off these dimensions, remembering that with three birds to a cage the width per bird is only $5\frac{1}{4}$ in. The only way in which these birds can stretch their necks is by putting them out through the bars of the cages.

Is it any wonder that these birds, and the broilers with their allowance of *up to* 0·8 *sq. ft.* per bird, have to be de-beaked to prevent them damaging each other? De-beaking involves the removal of up to half the bird's upper mandible with a machine designed for the purpose, but it may be done by anyone, experienced or inexperienced alike. This in itself presents a hazard to the bird. 'In inexpert hands the pain could be considerable,' an article in the *East Anglian Daily Times* 6th October 1962, points out. It goes on to cite a case where 'after the operation the bird seemed to hang its head because it was almost numb with pain'.

Bedichek comments on battery cages unequivocally in his book *Adventures with a Naturalist*:

> I have looked attentively at chickens raised in this fashion and to me they seem to be unhappy and in poor health. Their combs are dull and lifeless except for glaring and unnatural patches of colour that appear occasionally … the battery chickens I have observed seem to lose their minds about the time they would normally be weaned by their mothers and off in the weeds chasing grasshoppers on their own account. Yes, literally, actually, the battery becomes a gallinaceous madhouse. The eyes of these chickens through the bars gleam like those of maniacs. Let your hand get within reach and it receives a dozen vicious pecks – not the love peck or the tentative peck of idle curiosity bestowed by the normal chicken, but a peck that means business, a peck for flesh and blood, for which in their madness they are thirsting. They eat feathers out of each other's backs or, rather, pull out each other's feathers and nibble voraciously at the roots of the same for tiny blobs of flesh and blood that may adhere thereto.

The chief argument put forward by the industry for the contentment of the battery bird in its cage is the fact that they produce a high number of eggs in the year in which they are in the cage. But is it not true that every hen has a potential of eggs in her ovary to be laid during her lifetime? What the battery producer does is to ensure that all energy goes into laying the eggs, and moreover, into laying them as quickly as possible. The stimulation of the feed, of lighting patterns, and even of continuous light music in some units, is all aimed at egg production. Remember that the hen is only kept for a year before killing so that she must be made to lay to her maximum in that time.

Dr Knowles of the British Egg Marketing Board, has said that the important thing to remember in this unfortunate controversy is that the hen, *whatever the conditions*, will do her best to produce a first class egg. A corroboration of this came up in a discussion of cage layer fatigue when it was pointed out that the birds laid eggs until they dropped dead.

It was not known why they did not stop laying before this. Isn't this some indication of the inherent characteristic of egg laying, and also a guide to the amount of 'forcing' in their management? That she lays a large number of eggs is as shaky a criterion of happiness as that regarding the meat animal which is made to put on a lot of flesh in a given time. The battery bird also puts on weight – how can it help it? – but it is not healthy. Dr Blount himself has recorded that in one year alone an experimental unit recorded 'cancer of the heart, lungs, ovary, oviduct, kidney, leg muscles, liver and abdomen' (*Hen Batteries*). And cancer is only one disease from which these birds suffer. Mortality is generally accepted as being between 12 and 15 per cent, and up to 20 per cent including culls; one in five does not sound exactly thriving.

In Denmark battery cages have been legally abolished. An R.S.P.C.A. leaflet reports:

> According to a letter from the Danish S.P.C.A. in Copenhagen, the reasons which led their Government to pass such legislation were 'the many complaints from animal lovers in the country; hens were sitting penned up, shelved and never allowed out; their feet were damaged by wires in the bottom of the cages and they could not wear down their nails. The veterinary police and veterinary health authorities recommended the legislation proposal sent by the different societies (for the protection of animals) to the Ministry of Justice….'

It might be apposite to stress that Denmark is one of the largest egg producers in the world and has made a system of free range, or deep litter with access to free range in good weather, an economic and flourishing industry.

## Poultry Packing Stations

The poultry packing station would seem to flout all laws and all humanity in an effort to make a profit out of killing and processing broilers and battery birds.

The Conveyance of Live Poultry Order 1919, has the following clause to protect birds during conveyance by road or during exposure for sale; that the birds shall:

> 3.iii.  not unnecessarily be tied by the legs or be allowed to remain so tied for a longer period than is necessary; or unnecessarily be carried head downwards; …

I would point out again that the convenors of this Act knew nothing of poultry packing stations where, as can be seen in photographs in Figures 8 and 9 poultry are uncrated and suspended head downwards by their feet onto a conveyor belt. In some packing stations it is only a matter of seconds before the birds are stunned, but in others they can be suspended on the conveyor belt for anything up to five minutes before being either stunned or killed. Suspension on the conveyor belt allows a concentration of the chicken's blood into its head, and therefore a speedier bleeding when it is finally killed. It is considered a convenient way to kill the bird, and easier to handle when killing. But even if it is convenient and economic it is an acknowledgedly cruel way to handle a bird for long, and what is cruel in one place is cruel in another.

The Sussex D.P.K. (*Poultry World*, 12th July 1962) commented heatedly:

> I am surprised at the number of small poultry keepers who do not know the correct way of holding and carrying a fowl. Only yesterday, a youth of my acquaintance arrived at my back door, carrying a wildly flapping cockerel, suspended by the legs….
>
> I am a fairly placid individual, as a general rule, but this sort of thing always annoys me. It is so obviously a cruel and thoughtless method of handling poultry.

The Conveyance of Live Poultry Act also takes up the confinement of poultry in crates:

> **9.** Any person who in Great Britain, in connection with the conveyance of poultry belonging to him or being under his charge, shall cause or permit the same to be confined in a receptacle for a time longer than is reasonably necessary, shall be deemed guilty of an offence against the Act of 1894.

The owner of the packing station I visited told me that his vans went out at one o'clock in the morning to crate up the birds, which were then brought back and left at the side of the building awaiting the opening of the plant and then their turn on the conveyor belt. They could obviously quite easily spend a minimum of eight or nine hours being crated up before being released.

It is generally assumed that as the chicken is a creature of small intelligence it is uninfluenced by watching others of its species die and unable to anticipate its own death. How can we be sure of this? Has anyone done any actual tests to ascertain that this is a fact? Isn't approaching death an instinct with all animals whether of high or low intelligence? I noticed in the packing station that during the half minute when the live birds crossed the path of the dead birds they were apparently more frightened than at other times. Some of the crated birds had a grandstand view of the slaughterman at work for a considerable time before their turn came.

Now I come to that part of the 'processing' I feel to be most consciously and unnecessarily cruel.

As we have seen in the chapter describing conditions in a poultry packing station, Mr Wright of the Animal Health Trust watched birds going past a slaughterman without prior stunning and declared:

> … I have no hesitation in considering jugular severance without prior stunning as being grossly inhumane as birds were obviously fully conscious and in great pain for some appreciable time.

At the request of the Humane Slaughter Association, Mr Wright carried out further tests on heart beat and respiration on birds so slaughtered and found that two out of every five went into the scalding tank alive.

What I find so incredible is that the industry can quite happily plan increases in throughput per hour when it knows that killing without prior stunning is cruel and that no effective stunner has been invented to cope with the increased number. It can quite calmly and deliberately plan cruelty on a vast scale. Its excuse for this is commercial competition, and where commerce is concerned industry feels entitled to ignore cruelty.

Surely the packing station which has a stunner capable of dealing with its throughput of chickens, but which does not, for economic reasons, use the stunner, can be prosecuted for 'causing unnecessary suffering' even within the anomalies of our law? If this is not cruelty in law, one might almost ask what could be.

## Veal and Broiler Beef Calves

The R.S.P.C.A. is pressing for the blind beef calves to be taken direct from farm to slaughterhouse to avoid the horrors of the markets where 'some people dealing with them are not too kind'. Their Inspector for Hull, Mr Heath said:

> It is pathetic to see these animals in the market – they are in strange surroundings, collide with gates and walk into people.

But Mr Heath told the *Yorkshire Post* who reported this on 4th May 1963:

> No prosecution has been taken by the Society because it is difficult to prove there is substantial suffering within the meaning of the Protection of Animals Act – there does not appear to be any great pain in the eye – the animals just do not know where they are.

Surely the suffering of these animals lies in that they are blind. The cruelty inflicted on them, accidental though it may be, is in allowing them to become blind. A blind man does not necessarily have pain in his eyes. And why must we now require *substantial* suffering in order to prosecute? The words in the Act are *unnecessary* suffering.

From correspondence I have read in the scientific press about difficulties being experienced in rearing beef calves by quick-feed methods, it appears that a degree of pneumonia is almost inevitable in these calves. Coughs are an accepted thing in the units.

Blindness, coughs, damaged livers. Yet these animals with their feed additives of antibiotics, tranquillisers and hormones put on weight rapidly and are classed as 'quality' meat.

## Deficiency Diets

How far, within the meaning of the Act, could a diet causing deprivation of some substance vital to the health of the animal be construed as cruelty? Surely this is the *omission of any act* which causes unnecessary suffering. I would like to discuss this point with relation to both barley beef and veal calves.

The barley beef calf is so called because it is allowed only a concentrate consisting of bruised barley, fish meal and added vitamins (plus drugs) and no access to hay. The veal calf is even further deprived. It is fed entirely on a milk substitute. This is done deliberately so that the animal cannot become a ruminant when its flesh would darken. Both animals crave roughage.

A letter in *The Veterinary Record*, 22nd December 1962, gives the physiological explanation for this craving. It is written by Mr Brownlee, M.R.C.V.S., of the A.R.C. Field Station at Compton:

> The rumen 'desires' fill as indicated by: (*a*) calves on all-milk diet will consume bedding, even unpalatable peat-moss litter, which litter is present in the rumen in quite large quantity, (*b*) 27 calves, after 12 weeks feeding on a synthetic-milk diet and with no bedding allowed, all showed hair balls in the rumen (unrecorded observations) – I presume these hair balls do result from hair being actively retained as fill by the rumen – (*c*) the rumen does not empty on starvation (Colin, 1871).

Mr Brownlee then gives us his tentative conclusions:

> … (*a*) When a sufficiency of roughage is fed, any concentrate food will tend to leave the rumen relatively quickly, (*b*) where insufficient roughage is fed the 'desire' for fill on the part of the rumen will result in a measure of concentrates being retained leading to excessive fermentation and consequent lowering of pH….

Another veterinary surgeon has explained that 'though not proved, it is probably the development of toxins in the ingesta that causes the liver damage. The liver, of course, is the filter bed of the blood and any toxins which have been taken up into the blood stream will be concentrated in the liver and cause fatty degeneration'.

The blindness in the barley beef calves is due, it is suggested but not yet proved, to a shortage of vitamin A in their concentrate mixture. Four veterinary surgeons from the Cambridge University Department of Veterinary Clinical Studies, who have made a study of hypovitaminosis A in cattle, remarked:

> We are well aware, but still surprised, that many fattening compounds have no added vitamin A. There is evidence to suggest that the faster the growth of young animals the greater is their need for this vitamin and probably others too. (*The Veterinary Record*, 26th January 1963)

I have been unable to find a reason for the extreme shortage of vitamin A in the veal calf milk substitute. The British Veterinary Association booklet *The Husbandry and Diseases of Calves* suggests a minimum of 5,000 international units a day, and the English milk substitute analysis shows only 1,300 I.U., whereas the Dutch has only 200 I.U. This vitamin is an anti-infection safeguard and its absence has provenly caused blindness. It is cheap to include in the feed. Why on earth is so little included? The Minister states that any cruelty in these matters is covered by the Protection of Animals Act, 1911.

The veal calf milk substitute is *designed to produce anaemia* in the calves. The Ministry of Agriculture have admitted that on slaughter the haemoglobin level of the calves had been found to be between 5 and 7 g. per 100 ml. as opposed to the normal of 12 g. per 100 ml., indicating a very marked degree of anaemia. Why is it not punishable to rear animals on a food which is designed to make them unhealthy? Why does this not constitute an offence under the Act as an omission of an act which causes suffering? Surely the craving for roughage which is a physiological necessity to these calves, a deliberate withholding of sufficient iron which creates a need so strong that the calves will lick their own urine unless closely tethered, the prevention of rumination in the veal calf – a deprivation which prevents natural development and causes physiological changes in the animal – surely, we must admit that suffering is caused to the animal.

But the Minister states that any cruelty is covered by the Protection of Animals Act, 1911.

It is time we faced the facts and acknowledged the more subtle and insidious forms of suffering now being inflicted on our animals and carried to a degree which could not possibly have been envisaged by the enactors of the 1911 Act. It would seem to me to be taking our domination of the animal world beyond moral limits to cause ill health to an animal simply to produce pale flesh, the only attribute of which is the fulfilment of a snob requirement. Veal calf producers talk readily enough of cruelty, even admit that some of their methods are inhumane, but offer the flimsy excuse that they are only producing *what the public wants*. We must cease to pander to an unenlightened public. We have laws to punish perverted and ignorant children who torture animals because it gives them pleasure, it is time we applied these laws to causing suffering to animals because their carcases then are said to tickle our palates.

Another omission, not often mentioned in discussion of veal calf rearing, is that of water. I have shown that to stimulate a thirst strong enough to cause the animal to drink abnormal amounts of the milk substitute, *no water is allowed a calf even in a heatwave*. Because the calf then drinks abnormal amounts of the milk substitute it sweats, becomes thirsty again and at the next feed will again drink abnormally. This is splendid for the producer who thus causes the animal to get fat quickly, but not quite so splendid for the animal. A deliberate withholding of water must during a heatwave cause unnecessary suffering. But the Minister states that any cruelty in these matters is covered by the Protection of Animals Act, 1911. Further, complete lack of exercise causes the calves to suffer from bloat. Mr Williams of Boots' Farms explains the reason for this (*Farmer and Stockbreeder*, 30th January 1962):

> Company and exercise were good for calves. 'They must have a gallop,' insisted Mr Williams. His practice was to start running the calves together in larger pens as they began to take dry food. As soon as dry food got into the rumen, fermentation started, but a gallop released the gas.

Practically the only humane thing about a veal unit is that it is warm. Otherwise the veal calf's life is a short, acutely uncomfortable endurance test.

Calves cannot bear to be tethered and these calves are on such a short tether that they cannot turn their heads round. This practice is now spreading to the broiler beef industry. They are not even allowed enough room to lie naturally. A ten-day-old calf measured lying in the relaxed half moon shape natural to a calf took up every bit of 22 in. in width. These veal calves are allowed only 22 in. until they are twelve to fourteen weeks old, and this is made worse by the fact that they are abnormally fattened and are therefore relatively bigger at that age. The veal calf has to adapt itself to this confinement by tucking its feet under it instead of stretching its legs.

But the Minister states that any cruelty is covered by the Protection of Animals Act, 1911.

The calves are never cleaned. It is not considered economic. This is a source of discomfort to them made worse by the flies which buzz round their tails and about which they are powerless to do anything.

Finally there are still calves kept in the dark, or near darkness, in solid-sided crates, and bereft of the sense of being part of a herd which is so essential to their well being. It is worth emphasising here that all photographs one sees of veal calves have, of necessity, to be taken in the light, so that photographs do not give a true picture of the apathy and misery in which the calf spends its short existence. Only their bulging and strained eyes give an indication of their misery and lack of health.

Statements like the following from Sir John Hammond, Scientific Advisor to *Farming Express*, 28th July 1960, give a curious impression of the life of the Dutch veal calf:

> … Under this intensive rearing system they have luxurious well-fed lives for 12 weeks, which otherwise they would not enjoy. They lie and gorge themselves, growing fat and keeping happy.

Sir John had prefaced the above remarks with the following description of the actual conditions:

> The calves are kept in dark houses, without exercise and fed on a diet, lacking in iron, which makes for white flesh.

It is interesting to note that Holland, who originated these methods of veal production, has had representation from her own animal welfare societies for many years and at last, in September 1961, passed legislation ensuring that:

(a) between sunrise and sunset there should in the space provided prevail twilight to an extent in which the animals and their surroundings can be clearly distinguished, and

(b) the measurements of the space shall be such as to leave sufficient room for any animal to lie on each side without hindrance, to stand without hindrance and, when standing, to be able to move its head without hindrance.

I asked the Dutch Agricultural Attaché whether these regulations gave any indication of new sizes of pens and he replied:

> The new law for the Prevention of Cruelty to Animals does not give any obligatory sizes for calf pens. The Law does say that calves must be able to lie on both sides. The Institute for Agricultural Farm Buildings does advise people to build pens with a width of 60 cm. (24 inches), whereas the length should be 1·5 metre (about 5 feet); the 0·5 metre ($1\frac{1}{2}$ feet) at the rear does not have a partition. The height of these partitions is 1 metre (about 3 feet).

No legislation was made regarding shortage of iron in the calf's diet, also a source of concern to animal welfare societies in Holland. Presumably commercial interests were too much at stake to allow for legislation in this.

It appears, however, according to a report in *The Times*, 15th August 1960, that:

> The Dutch domestic market … is less insistent on white meat than some of the countries to which Dutch veal is exported.

The Dutch recognise that whiteness of flesh is a fad rather than a characteristic of quality, and are prepared to fulfil this only for the export trade.


## Pigs

The plight of intensively reared pigs in overcrowded pens containing no bedding, in dim light or in darkness, usually feeding off the floor and with no separate dunging space, is made all the more acute by the fact that pigs are naturally clean, lively and intelligent animals.

It is commonly put forward that pigs are descended from a tropical species and should therefore be used to darkness and humidity, these being the conditions of the jungle. But in fact a more likely assumption is that put forward by Claud Miller, in *Animal Life*, June 1963:

> It is probable, moreover, that all the domestic pigs of today are predominantly descended from *Sus scrofa*, the wild boar of Asia and Europe.

This would mean that our pigs are inherently used to a temperate climate, and this inherent disposition has been intensified over hundreds of years of domestication.

Claud Miller explains that the pig's joy of wallowing in the mud is not because it is a dirty animal by nature, far from it, but because:

> ... the pig's sweat glands are inadequate, and, further, the thick layer of fat in which it is enveloped acts as an insulator and interferes with the transmission of temperature changes between the interior of the animal's body and the surrounding atmosphere.
>
> When the weather is very hot the pig therefore has great difficulty in keeping its blood temperature low enough for comfort, so it seeks patches of shade, access to the air, and if this does not help, if there is a puddle within reach, it rolls in it to moisten its skin.

In other words, a dirty pig is a reflection on its stockman for not providing it with adequate shade.

The 'sweat-box' piggery takes no account at all of the pig's physiological make-up. By careful insulation, tight packing of pigs, and no forced ventilation, a temperature of around 80°F. can easily be maintained, with a relative humidity of 90 to 96 per cent. 'This latter figure means it was practically raining,' says an article in *Farmer and Stockbreeder Supplement*, 13th March 1962, and goes on further to describe the atmosphere of the piggery:

> The heat was oppressive and the atmosphere close; the floors were often wet with urine and dung; walls and roofs dripped with condensation; and moulds and fungi flourished.

It is apparent that, with the pig's inadequate sweat glands and its consequent inability to get rid of body heat, life in these piggeries must be acutely uncomfortable and miserable. Indeed all the photographs one sees of the pigs are of an inert, closely packed mass of pigs lying on the floor. A more practical proof of their misery is that this same article says that 'hothouse pigs tend to dung and urinate over their lying area and even in the troughs'.

But the Minister states that any cruelty is covered by the Protection of Animals Act, 1911.

Regrettably for the pigs, they do live and put on weight, although they do it more slowly because they eat less. The *Farmer and Stockbreeder* article tells how in one unit 'production has "exceeded all expectations" ' and then goes on to suggest a reason why these animals do not succumb to disease:

> ... The higher the density of pigs, the greater the amount of heat produced on the floor and, since hot air rises and finds its way to the outlets, the higher the rate of ventilation; and the increased number of air changes per hour decreases the number of bacteria in the building.

> But it is also partly due to the high humidity which causes the 'sedimentation' of dust and, with it, of dust-borne bacteria. Thus many of the enemies of health are either carried away on the breeze or literally beaten to the ground.

This theory was put forward by Dr Gordon of the Northern Ireland Ministry of Agriculture Veterinary Department after close study of the system and has also been discussed by Dr Sainsbury of Cambridge University Veterinary School.

We have still to discover the long-term effects of using these methods. Meanwhile they are said to make money for the producers, so their popularity is increasing and they have spread from their original home in Northern Ireland to many parts of England.

Before I leave the question of pigs, I would like to mention one last small piece of research done by the chief pig officer of a feeding firm. He took a recording of the grunt a sow makes when she wants her piglets to suckle and reported with great pride that he had discovered that by playing it off at night-time he could make them suckle more often and consequently put on weight more quickly.

I have mentioned some cruelties inflicted on some animals, but intensive rearing is practised over a much wider field and it can be taken for granted that the inhumanities of overcrowding, lack of exercise, boredom, darkness and slats or wire, can be applied to every animal the sole function of which is to 'convert food to flesh'.

I would like finally to mention a trend which has great potential dangers for the animal, the advent of a five-day week for farm labour. *Farming Express*, 10th August 1961, quotes a farm where this system is already in operation:

> So now the 30,000 layers, 8,000 growers and 3,500 head of breeding stock … are left from mid-day Saturday until Monday morning without attention. And every member of the staff works a five or $5\frac{1}{2}$ day week.

And *Farmer and Stockbreeder*, 4th April 1961, reported:

> Beef production on a five-day week: this revolutionary concept was presented … by Dr T. R. Preston of the Rowett Research Institute … (who) envisaged giant beef 'factories', with hoppers self-feeding animals on concentrates, operated by time switches, which would do away with the need for weekend work.

What happens if the machinery breaks down? What happens if an animal falls ill on Friday evening? It could exist in agony until Monday morning. Surely it is as immoral to leave an animal, especially in the conditions we have discussed, alone and untended for two days and three nights, as it would be to leave a child, even supposing the child had sufficient food and drink for that time? This is a move which should be checked before it has had time to gain ground and become accepted along with all the other niceties of the system.

In 1961 Mr John Dugdale, supported by Mr Burden, Mr Anthony Greenwood, Dr King, Sir Thomas Moore, Mr Moyle and Mr Russell, put forward a Private Members' Bill, called the 'Animals (Control of Intensified Methods of Food Production) Bill', 'To authorise the Minister of Agriculture, Fisheries and Food and the Secretary of State for Scotland to make regulations for securing humane conditions and practices in connection with the rearing and keeping in buildings of animals for the production of food, and the slaughter of such animals; and for purposes connected therewith.' This Bill did not have government support and never had a second hearing, and has not managed to get a hearing since.

R. Trow-Smith, writing in *Farmer and Stockbreeder*, 13th September 1960, said:

> … In this business it is all too easy to let a little inhumanity in at one end so that a little more profit may come out at the other.

# Conclusion

The arguments against factory farming are essentially based on humanitarianism and quality; the arguments for factory farming, such as they are, are economic arguments. We need not be disturbed about this. Throughout our society there are clashes between economic and social considerations; this has been true from the first factory legislation onwards. Our legislation is full of examples of laws which for humanitarian reasons prevent people using the cheapest methods. So, to set against the obvious humanitarian reasons for not adopting factory farming, we must look at the strength of the economic arguments in favour.

These take two forms, first that factory farming is cheaper, that must mean that it is economically more efficient; and secondly, that the world's need is for more food, and that this is the best way to increase output.

We will look at these two arguments in turn. Let us start with the economic argument as applicable to the individual producer.

Labour is scarce and expensive, intensivism utilises what labour there is to the greatest advantage thereby reducing costs to a minimum. Buildings are expensive so as many animals must be crowded in as is possible.

In the chapter on battery birds I have quoted figures which show that the capital cost of the buildings and battery cage equipment combined represents only 8 per cent of the cost of producing eggs, and as this is the most sophisticated and expensive equipment used in poultry rearing, simpler housing will represent less than 8 per cent of the cost of poultry production.

 © J. Harrison and J. Wilson 2013. *Animal Machines* (Ruth Harrison)

Comparable figures can be arrived at for each type of stock. To take pigs as an example. The figures of cost in the table opposite are taken from *Farmer and Stockbreeder*, 13 March 1962:

|  |  |  |  | per cent | per cent |
|---|---|---|---|---|---|
| Cost of purchasing 70-lb. weaners | £7 | 15s. | 0d. | 60·8 | — |
| Feeding | £4 | 10s. | 0d. | 36·0 | 91·6 |
| Labour |  | 3s. | 4d. | 1·3 | 3·5 |
| Housing |  | 3s. | 4d. | 1·3 | 3·5 |
| Overheads |  | 1s. | 6d. | 0·6 | 1·4 |
|  | £12 | 13s. | 2d. | 100·0 | 100·0 |
| Average sale price of pig: | £16 | 10s. | 0d. |  |  |
| Profit per pig: | £3 | 16s. | 10d. |  |  |

These figures are even more startling than those quoted for poultry. Of the overall capital required to run a pig unit the cost of the buildings to accommodate the pigs represents only 1·3 per cent. Of the running costs of the establishment, deleting the cost of buying in the weaners, which will have borne their own proportion of the respective outlays, it represents only 3·5 per cent. Of the profit per pig it represents only 4·3 per cent.

Lest the above figures be challenged as unrealistic let us get at the answer in another way.

| *Cost of building:* | *£1 per sq. ft.* | | | *£2 per sq. ft.* | | | *£3 per sq. ft.* | | |
|---|---|---|---|---|---|---|---|---|---|
| Cost per pig place at: | | | | | | | | | |
| 3½ sq. ft. per pig | £3 | 10s. | 0d. | £7 | 0s. | 0d. | £10 | 10s. | 0d. |
| 5 sq. ft. per pig | £5 | 0s. | 0d. | £10 | 0s. | 0d. | £15 | 0s. | 0d. |
| 10 sq. ft. per pig | £10 | 0s. | 0d. | £20 | 0s. | 0d. | £30 | 0s. | 0d. |

Throughput of 50 pigs during life of building (12½ years)

| Cost per pig: | | | |
|---|---|---|---|
| 3½ sq. ft. | 1·4/- | 2·8/- | 4·2/- |
| 5 sq. ft. | 2/- | 4/- | 6/- |
| 10 sq. ft. | 4/- | 8/- | 12/- |

Throughput of 100 pigs during life of building (25 years)

| Cost per pig: | | | |
|---|---|---|---|
| 3½ sq. ft. | 0·7/- | 1·4/- | 2·1/- |
| 5 sq. ft. | 1/- | 2/- | 3/- |
| 10 sq. ft. | 2/- | 4/- | 6/- |

To these figures must be added interest on capital and maintenance, but these two items combined should not exceed 10 per cent of the above costs.

£3 per sq. ft. is a lot to pay for a piggery and compares with the figure of £4 per sq. ft. for which we can currently build schools in this country, and £2 10s. 0d. per sq. ft. for which we build houses. The more realistic figures are those of the first and second columns.

These figures confirm the previous example and illustrate that the overall economy to be gained by increasing the intensity of stocking from, for example, 10 sq. ft. down to 5 sq. ft., and particularly from 5 sq. ft. down to $3\frac{1}{2}$ sq. ft., is in itself quite marginal in relation to the overall cost of production, except in buildings costing £3 per sq. ft. amortised over the short period of $12\frac{1}{2}$ years, when it becomes significant.

The point that glares most obviously from these cost figures is that labour and housing between them are minor items in rearing costs and there is no economically viable argument for the unreasonable crowding of animals into the buildings. Overwhelmingly the greatest cost in rearing these animals is their food.

With regard to labour it must be stressed that all the weight of research towards labour saving has been concentrated on intensifying indoor stock. Hardly ever have I found evidence that any concentrated effort is being applied to economise on labour for stock rearing out of doors. Surely it is not beyond the ingenuity of science to achieve comparable results in more natural methods of stock rearing?

To take some examples. The automatic feeding and watering systems which have been devised for indoor stock and which are the greatest savers of labour could just as easily be adapted for use for outdoor stock. Moreover it is becoming increasingly evident that the labour which it was thought would be saved is often absorbed in other ways; for example, in the case of poultry in laborious de-beaking and repetitive vaccination operations, in the case of all intensively kept animals a higher incidence of disease must mean more time being consumed by veterinary officers and others in dealing with it, and if the cost of all the ancillary services were to be analysed the labour content in the production of buildings, mechanical appliances, feed, drugs and so on might well prove to have outweighed the saving in agricultural labour. One begins to wonder whether it might not be cheaper to eliminate the labour shortage in this field by raising the status of the agricultural worker.

The overall economic argument for intensive rearing of animals in this country goes somewhat as follows: we are a small island with a large population and insufficient land to sustain us all in food. We must, therefore, produce as much as we can from our land. With these new methods of animal rearing we can boost our production by taking animals off the land and feeding them as necessary with imported feeding stuffs. The land so vacated can be turned over to other crops, thereby increasing our overall food production.

The whole argument rests on our ability to re-use the land vacated by stock in an equally adequate manner. Is this being done? What is being produced on this freed ground?

The insular conception of self sufficiency implied in the above argument is, however, out of date, and since it is only sustained at the expense of a subsidy to the tune of some £340,000,000 a year it would appear to be as expensive as it is unsound. It seems undesirable to boost production which can only be sustained by subsidy of this order, for it means that the more we produce the more we pay. Instead of striving to increase subsidised production it would be sounder economics to import cheaper food from abroad and export other goods to pay for it.

Agricultural support now amounts to 2s. 6d. per week per head of population, or round about 10s. per week for an average family, against a food bill which excludes exotics such as imported fruit.

Subsidies are a philanthropic measure and they are there, if they must be there at all, to benefit the nation and ultimately each one of us as individuals. But we have seen that they are being abused by essentially short-term argument to produce an unnecessary quantity of inferior food. Surely the less land we have the more important it becomes to concentrate on getting the maximum amount of *quality* food from it. We cannot afford to waste any of our production capacity on food of a lowered food value and must safeguard every scrap we produce. The whole basis of subsidy should therefore be reassessed so as to encourage quality rather than indiscriminate quantity.

The industry might, however, feel that its efforts have been justified, for it has produced results which the public can see, the Rotissomat, the veal cutlets, the cheap and abundant supply of chickens and eggs, and in their innocence the public has taken advantage of these increases. Weekly consumption per head has risen, of eggs from 3·46 in 1950 to 4·64 in 1960, and of poultry from 0·35 ounces in 1950 to 1·68 ounces in 1960. We are indeed one of the most lavishly fed countries in the world. But what good does this do us if the food is of an inferior quality?

'We have brought the price of a chicken down from 15s. to 8s. 6d.', the industry is continually trying to impress on us. Can we be certain that the product is cheaper? It may look the same to the ordinary consumer but in fact it is much less nutritious.

One of the major benefits the country has conferred on agriculture is a total exemption from rating. Until recently factory farmers had taken for granted that this benefit would extend to them, but local authorities are waking up to the fact that factory farms are really industrial concerns and recognising that these farmers are forfeiting their rights to derating under the Rating and Valuation (Apportionment) Act 1928, which allowed for buildings used in conjunction with agricultural practice on the land to be exempt from rates, for example a barn used for housing hay or machinery. A factory farm which is independent of the land in that the animals never set foot on it, are fed from manufactured foods, and where even the dung is not spread back on it, cannot be counted as an agricultural practice and is therefore liable to rating along with any other industrial premises.

Factory farming would seem to have put us into a ludicrous situation with regard to fertiliser subsidies. We import £20,000,000 worth of fertilisers each year and pay out £35,000,000 on subsidies, whilst at the same time because of intensive rearing methods so much farmyard manure is fed down public sewers that local authorities are having to impose quite a heavy charge to cover the costs of dealing with it and rendering it valueless. It has been said that one hen produces 1 cwt. of droppings a year and that one ton of droppings an acre is all that is needed for fertility of the soil. Twenty birds to fertilise one acre. I cannot improve on the argument of this letter from *Poultry World* (19th July 1962):

> Subject of this letter is poultry manure, or rather the waste of it.
>
> Many millions of tons are being disposed of annually – deep pits, ravines, town dumps, and by means of various methods of incineration, etc. There have even been cases of sewage disposal plants being clogged with it.
>
> Incredible – but this 'waste product' is the most valuable farm land fertiliser imaginable.
>
> When I trebled my laying cages last year, I had an opening cut in the wall at the receiving end of the mechanical cleaning device to lead into a two-bay manure shelter. As each bay became full I distributed the manure by means of the usual tractor-drawn dung spreader over my grazings.
>
> The result has been spectacular. I am already carrying 70 per cent more cattle on the same acreage. My milk yield per cow has gone up; my hay crop has doubled, and whereas cows avoid cattle manure areas until it is weathered by time and frost, they actually follow closely behind poultry manure which sweetens the ground.
>
> My hay is almost in, and the result is as near double last year's as doesn't matter, in spite of a very serious drought in this locality.
>
> On various occasions I have brought this matter up at farmers' meetings and in every case there has been a shriek of 'We don't want to handle that stinking stuff.'
>
> Poultry dung is organic as against chemicals, if there *are* any valid grounds for preference.
>
> I would go so far as to suggest that no farmer be allowed to use *subsidised* fertiliser unless he can prove that poultry manure is geographically unobtainable. The benefit to the hen keeper would be very great because by degrees, instead of having the costly and worrying disposal problem before him, he would in due course be able to sell what is now a liability.
>
> I recently had a word with my farm contractor on this matter. He is a top line efficiency general farmer of the middle bracket acreage group. He said, 'I keep a lot of poultry, and as long as at least I can break even on them I shall continue, as the value is in the manure spread over my land.'

It boils down as an economic argument to this summary: a case can be made for intensive rearing if the food produced is thereby cheaper and if the end product is the same as the products of traditional farming. Inasmuch as the end product is inferior the economic argument is vitiated.

Now to take up the second point with which we began this chapter. Intensive rearing of animals is said to be aimed at increasing the animal protein content of our food and meeting the requirements of an ever increasing population. This argument has no force in the western world since nutritionists are generally agreed that too many animal calories are already being consumed. Some of the industry's more hopeful protagonists would have us believe that these methods might go some way towards the solution of the food problems in underdeveloped countries. Today's world population of 3,000 millions is expected to double by the end of the century and, according to a U.N. survey, to consist of 1,352 millions in high calorie countries and 5,089 millions in low calorie countries. The onus of helping the low calorie, or underdeveloped countries falls naturally on the more affluent and technically advanced countries of the west.

Although, where practicable, it has been found helpful to send surplus food, taken off the markets of the west to ensure stable prices to the producers, to underdeveloped countries as a short-term relief measure, this cannot be regarded seriously as a long-term policy. The World Poultry Congress of 1962 recognised the economically unsound conception of any long-term disposal of surpluses of poultry and eggs to the East. *Poultry World* reported:

> It is clear that few of the delegates have believed that 'scalping' surplus from over-producing countries and sending it to those suffering food shortages is anything other than a temporary palliative of dubious value.
>
> Apart from the question of payment it must be obvious that the quantity of eggs or poultry meat from such a plan would be too small to be in any way effective. As our delegates pointed out – a 5 per cent increase in British egg production would give only one egg per year to everyone in China.
>
> Furthermore if the leading poultry countries became geared to this kind of relief work any ultimate improvement in production achieved in the under-developed countries would throw the former into a state of economic chaos.
>
> To those of us here who recall the effect on egg prices that only one half per cent increase in the supply of imported eggs had upon the British market this needs no emphasis. (16th August 1962)

The world food problem is of such magnitude and urgency that it must be met by the quickest and most economical means available. Five to ten times as much vegetable protein can be produced per acre as animal protein, and the experts realise that this is the first answer to the problem. Ultimately the best help we can give the undernourished is to help them to help themselves, to increase the scope and productivity of their natural resources, and further to teach them to use wisely those resources they have. You may well feel that this is a case of the halt leading the blind. It is to be hoped that care will be taken to avoid wasteful methods we have in the West, firstly of processing much of the natural goodness out of the food, and then the inclusion of additives, few of which have been tested thoroughly for safety.

F.A.O., seventy years after the discovery of Vitamin B, are still trying to educate the peoples of the East that the highly-polished white rice which they feel to be socially superior to their unmilled rice is inferior in that all the essential thiamine is processed away, thus spreading the scourge of beri beri, just as we in the West still need educating on the subject of our devitalised white bread.

Protein deficiency is even more acute than vitamin deficiency, and is especially tragic in the East in that it causes high child mortality in an illness called 'kwashiorkor', or 'the sickness the child develops when another baby is born'. A practical and highly successful means of producing milk from vegetables and seeds has been evolved and this will go a long way to meeting the tragic problem of the child who starves after being taken off its mother's breast. Another means, also of exploiting natural resources to maximum advantage, has been developed by N. W. Pirie, F.R.S., of Rothamsted Experimental Station. See bibliography. This is a leaf protein which is actually being processed and used in India and elsewhere. These processes can utilise what would otherwise be wasted. For example, Mr Pirie states that the residues from oil seeds, after the oil has been extracted, 'if properly conserved and distributed could satisfy about a third of the world's protein needs'.

A *Financial Times* survey on Science and Food (19th September 1962) points out more examples of foods which have hitherto not been sufficiently explored by man:

> Yeast and certain algae are examples. These are edible and high in protein content and, moreover, can be cultivated speedily; 200 tons of yeast, which has a food value, in terms of protein, equal to that of 500 bullocks, can be produced by a large factory in one week....

Richie Calder has said, in *The Sunday Times* (17th March, 1963):

> It will need the combined intelligence, and efforts, of mankind to increase yields, recover the deserts, extend cultivation into climatically inhospitable regions and to farm the oceans which cover seven-tenths of our globe.
>
> All this is possible – without even a fresh recourse to science or an unforseeable breakthrough....

## Recommendations

I have already made the suggestion that the emphasis of subsidies should be on quality rather than quantity. What further measures can be initiated to protect both human beings and animals from the undesirable practices I have discussed in this book.

Three separate measures are required to protect the health of the people:

**1.** The safeguarding of our food from soil to plate from additives the effects of which are not thoroughly tested and understood.
**2.** The consumer must be provided with the information to allow him the option of choice in his food.
**3.** A reassessment of true quality in food.

To implement the first of these measures we need in this country a counterpart of the American Delaney Committee. As Bicknell points out, 'any of about 1,000 alien chemicals are now deliberately added to food in this country or abroad. The number increases. With a few exceptions any chemical may be used in food without telling the consumer or health authorities, and without previous investigations on its toxicity or its effects when eaten for many years.' The sole function of this Committee would therefore be to investigate all additives and safeguard our food from those the effects of which are not thoroughly tested and understood. This would seem to be necessary in addition to the new Consumer Council which has just been set up in that that body's terms of reference are so broad that they could not possibly hope adequately to cover this specific subject with the thoroughness it requires, on top of all the other demands which will be made on them. It might of course be found convenient to implement this measure through a sub-committee of the Consumer Council, providing it could be vested with sufficient authority.

The second measure requires new legislation. We have already in this country legislation requiring the disclosure of the constituents of patent medicines and drugs offered for public sale. This disclosure must be extended to cover food. In America every single constituent must be declared on the label of a packet or tin of food. In Germany this cover was made even more all-embracing. In the spring of 1961 all the women members of the Bundestag (German House of Commons) joined together and got laws passed compelling the disclosure of every non-nutritive additive and adulteration of food. The slogan 'The Truth on the Label' got taken up all over Western Germany. Thus, for example, the dye in butter and cheese had to be declared, the spray on apples; tinned vegetables could not be dyed. Even restaurants had to declare colourings or flavouring agents on their menus. The result of these laws was to make it so complicated for retailers and restaurateurs that naturally grown foods were highly sought after and fresh sauces and flavourings made on the spot to save the addition of these disclosable ingredients. This, it might be assumed, is one of our rights as consumers, that we know what we are eating. If we must shoulder the risk, we must be allowed the choice. Not a bit of it. The Egg Marketing Board have complained of too much labour involved in separating free range from intensively produced eggs. Butchers do not really seem to know which of their meat has been intensively reared or has had hormones added.

The third measure is not a matter for legislation but for education. This is the fundamental problem. If our knowledge of nutritional values was adequate and widespread the first two measures, which are by way of being palliatives, would hardly be required.

The prime need for the protection of animals against these intensive forms of rearing is a new Protection of Animals Act. Mr Dugdale's Bill sought to control the conditions, but in my opinion it implied too great an acceptance of conditions as they now exist. The recent Dutch measures for the protection of veal calves also fall short inasmuch as they also accept the existence of extreme methods as a fait accompli.

Included in a new charter for animal welfare I would like to see:

**1.** The complete abolition of battery cages for laying hens.
**2.** The complete abolition of the intensive methods now used in veal production. Both these practices are as unnecessary as they are undesirable.

The abolition of egg and veal production by these methods in this country would have to be reinforced by the abolition of *imports* of eggs and veal produced by corresponding methods. It would be a poor gesture to ban the methods in our own country whilst supporting by imports the same methods used elsewhere.

**3.** I would like to see specific legislation banning the rearing of animals on deficiency diets. This would preclude food designed to produce anaemia such as is fed to veal calves, and the possibility of blindness in barley beef calves.
**4.** Permanent tethering should be banned.
**5.** Slats should be banned.
**6.** The keeping of animals in dim light or darkness should be banned. This is a sign of bad husbandry and is completely unnecessary where conditions are satisfactory for the animals.

Legislation alone will not provide the animals with an adequate charter. We need to reassess our basic attitude towards the animals which are bred solely for human benefit. Here again education is needed throughout the whole fabric of our society.

Assurances that all is well in the world of farm animals have not been lacking in the immediate past and will certainly not be lacking in the future. In these present times, when the pressures of intensive living bear weightily on our shoulders in many fields, the professional and the official reassurer are ever with us. We shall be assured that no cruelty is involved in intensive rearing and that, if it is, the Protection of Animals Act 1911 will adequately take care of it. We shall be assured that the products of the industry are better and more nutritious than they have ever been and we shall be told that we are the best fed people on earth and that we are becoming better fed every day. It will be imputed that those with a sneaking suspicion that all is not well are a decided minority, that they are faddists, and that those who hold more extremist views on the subject are cranks. The apparatus for reassurance is formidable. Bland faces on our television screens will constantly be appearing, backed by weighty qualifications and disarming smiles, to pour out reassuring bromides on any and every aspect that may be causing disquiet. Against this I can only set down the facts as I see them and rely upon my reader to form his own conclusions.

# Bibliography

## Books

E.B. BALFOUR, *The Living Soil*, Faber and Faber, 1959.

ROY BEDICHEK, *Adventures with a Naturalist*, Gollancz, 1948.

DR FRANKLIN BICKNELL, *Chemicals in Food and in Farm Produce: Their Harmful Effects*, Faber and Faber, 1960.

W.P. BLOUNT, *Hen Batteries*, Baillière, Tindall and Cox, 1951.

RACHEL CARSON, *Silent Spring*, Hamish Hamilton, 1963.

LEWIS HERBER, *Our Synthetic Environment*, Jonathan Cape, 1963.

JORIAN JENKS, *The Stuff Man's Made Of*, Faber and Faber, 1959.

J.O.L. KING, *Veterinary Dietetics, A Manual of Nutrition in Relation to Disease in Animals*, Baillière, Tindall and Cox, 1961.

R.G. LINTON AND GRAHAME WILLIAMSON, *Animal Nutrition and Veterinary Dietetics*, W. Green and Son Ltd., 1943.

KONRAD LORENZ, *King Solomon's Ring*, Methuen and Co. Ltd., 1952.

SIR ROBERT MCCARRISON and H. M. SINCLAIR, *Nutrition and Health*, Faber and Faber, 1961.

H.J. MASSINGHAM, *England and the Farmer*, Batsford, 1941.

MICHEL PERRIN, *Essai de Caracterisation des Viands de Veau Insuffisantes*, Foulon, 1953.

L.J. PICTON, *Thoughts on Feeding*, Faber and Faber.

WESTON PRICE, D.D.S., *Nutrition and Physical Degeneration – A Comparison of Primitive and Modern Diets and their Effects*, The American Academy of Applied Nutrition, Los Angeles, 1950.

Rural-Reconstruction Association, *Feeding the Fifty Million*, Hollis and Carter, 1955.

*Scientific Principles of Feeding Farm Livestock*, Proceedings of a Conference held at Brighton, 1958, Farmer and Stockbreeder Publications Ltd., 1959.

*Second Conference on the Health of Executives 1960*, The Chest and Heart Association.

LEONARD WICKENDEN, *Our Daily Poison*, The Devin-Adair Company, 1956.

G.T. WRENCH, M.D.(LOND.), *The Wheel of Health – A Study of a Very Healthy People*, C. W. Daniel Company Ltd., 1946.


## BOOKLETS AND PAPERS

Royal Institute of Public Health and Hygiene. Papers given at Public Health Conference 1962.

Soil Association, *The Haughley Experiment 1938–62*.

The British Veterinary Association, *The Husbandry and Diseases of Calves*.

Proceedings of the Vitalstoffe-Zivilisationskrankheiten steht unter dem Protektorat des Wissenschaftlichen Rates der Internationalen Gesellschaft für Nahrungs- und Vitalstoff-Forschung.

Universities Federation for Animal Welfare, *Courier*, Autumn 1960; and *The Scientific Basis of Kindness to Animals*, by John R. Baker, 1955.

Freedom from Hunger Campaign, *Third World Food Survey*, Basic Study No. 11, and *Malnutrition and Disease*, Basic Study No. 12, World Health Organisation, 1963.

Ministry of Agriculture publications, published by HMSO:

*Report of the Committee on Fowl Pest Policy*, 1962.

*Beef Production*, Bulletin No. 178.

*Poultry on the General Farm*, Bulletin No. 8.

*Poultry Housing*, Bulletin No. 56.

*Intensive Poultry Management*, Bulletin No. 152.

*Table Chickens*, Bulletin No. 168.

*Hybrid Chickens*, Bulletin No. 180.

*Rations for Livestock*, Bulletin No. 48.

*Calf Rearing*, Bulletin No. 10.

*Experimental Husbandry*, Nos. 1, 2, 3, 4, 5, 6.

*Experimental Progress Report*, 1961

*Incubation and Hatchery Practice*, Bulletin No. 148.

*Modern Rabbit Keeping*, Bulletin No. 50.

*Housing the Pig*, Bulletin No. 160.

Ministry of Agriculture Advisory Leaflets.

Poultry World Publication: *Egg Productions in Laying Cages, Fundamentals of Nutrition*, Nos. 1–8, by Dr Frank Wokes and Cyril Vesey, reprinted from Good Health.

DR N.W. PIRIE, *A Biochemical Approach to World Nutrition*. May and Baker Laboratory Bulletin, May 1961.

# Index

advertising  36, 80, 104, 151
Alp, Dr H. H.  168
American Cancer Society  149
American Food and Drug
        Administration  157, 159
Animal Health Trust  8, 59, 112, 152,
        154, 166, 185
'Animals (Control of Intensified
        Methods of Food Production)
        Bill'  193
antibiotics, for disease suppression
        and as growth stimulants
        38, 135
Armstrong, D. H.  38
arsenic as a growth stimulant  157
        in America  157
arteriosclerosis  171, 172


Baird, Eric  71
Baker, Dr John  73, 206
Bakker, Dr T. J.  87, 95, 97, 103, 104
Bartlett, Dr Stephen  181
battery hens *see* poultry
Beacon Milling Company  73
Bedichek, Roy  151, 176, 177, 183
beef calves *see* calves

Beeston, Norman  54
Bellis, David  152
Berglas, M.  149
Bicknell, Dr Franklin  155–158, 166,
        172, 201
Biskind, Dr Morton  165
Blount, Dr W. P.  68, 77, 164,
        169, 181, 184
boredom in animals  12, 37, 118, 177, 192
broiler chickens *see* poultry
Brownlee, A.  187


Calder, Ritchie  200
California, University of  62
calves, herd animals  104
        American feed lots  107
        beef  186, 187, 202
        blindness  186–188, 202
        deficiency diets  202
        prevention of rumination  188
calves, veal, anaemia  92–105, 188, 202,
        in Denmark  105
        deprivation of iron
                94–103, 105
        deprivation of vitamin
                B12  94, 99, 102

calves, veal, anaemia (*continued*)
      deprivation of vitamin A
          98, 99, 102
      deprivation of water 94, 95,
          97–99, 102, 103
      hair balls 91
      methods in Holland 102–105
      suckling and urine licking 93
      sweating 92
      tethering 91–93
      whiteness of flesh 105, 190
      *see also* milk substitutes
Cambridge University Department of
      Veterinary Clinical Studies 187
cancer in broilers 157
      in man 157
Carson, Rachel 3, 5, 149, 150, 157, 164–166
Cherrington, John 107
chloramphenicol *see* antibiotics,
      growth stimulants
Clapham, Robin 61, 169
Clark, J. 147
coccidiostats 46, 156
Coles, Dr Rupert 73, 82
Conveyance of Live Poultry Order,
      1919 184, 185
Coonoor, experiments with rats *see*
      McCarrison
Cotton, John 59, 60
County Landowners Association 108

Danish S.P.C.A. 61, 111, 184
darkness in intensive houses 35, 44, 67,
      87, 89, 104, 151, 177, 180, 189, 202
DDT 164–168
deficiency diets *see* calves
degenerative diseases 39, 149
Delaney Committee 160, 161, 201
ducks 58, 106
Dugdale, John 193
Dutch Institute for Animal Husbandry
      and Meat Production 104
dyes *see* eggs

Easterbrook, Laurence 93, 107, 170, 182
eggs, internal quality 79
      in America 76

      in Australia 78
      smallness of yolks 79
      vitamin A and D content 80
      yellow dye 78
Enders, Dr Robert 160–162
exercise 12, 37, 75, 83, 91, 92, 152,
      175, 189, 192
expectancy of life 149

F.A.O. 200
*Farmer and Stockbreeder* Vet. 93, 153,
      154, 156, 178–180
five day week for farmers 192
flies, as pests 164
      in America 164
Frazer, Professor A. C. 156
Fry, N. L. E. 5, 57
frustration of instincts 37, 77
fungal infections 156
Fussell, Dr M. H. 152

Gammans, Lady 97
German food laws 201
Gordon, Dr, of Belfast Ministry of
      Agriculture 192
Gordon, Dr R. F. 152
growth stimulants *see also* antibiotics,
      arsenic, hormones,
      chloramphenicol

Hailsham, Lord 167
Hale, Murray 57, 61
Halpin, Professor J. 68
Hammond, Sir John 189
Hartman, Dr Carl G. 160, 161
Haughley experiment 146, 147
hazards of intensive rearing 169
Herber, Lewis 154–157
hormones as growth stimulants 38,
      103, 135
      in America 157
      in Australia 78, 156
      in Italy 87, 163
Howard, Sir Albert 145
Humane Slaughter Association
      59, 60, 186

Hunza race  143, 145
Hutt, Dr F. B.  78–80

Iglesias, Rigoberto  162
immobility of animals  35, 118
inbreeding  75
insecticides and pesticides  149
          in New Zealand  149

Jennings, Sydney  33, 95, 96
Jones, Dr Ron  18, 71

Kennedy, H. R. C.  75, 76
King, J. O. L.  100
Knight, Granville F.  162
Knowles, Dr  183

Lambs  106, 162
Lorenz, Dr Konrad  48, 57

Manchester Corporation Act  63
Martin, W. Coda  162
Max Planck Institute  48, 152
McCarrison, Sir Robert  40, 141–143
Medical and Panel Committees of the
          County Palatine of Cheshire  142
medical training  140
milk substitutes for veal calves  87, 88,
          91, 94, 96–103
Miller, Claud  38, 190, 191
Milton, Dr Reginald  102, 168,
          170, 172
Missouri, University of  73
Moore, A. C.  77, 79, 100, 121, 135
Muller, Professor E. A.  152
myoglobin  92 see also calves

National Health Service  142
National Institute for Research in
          Dairying  91, 101, 102, 179, 181
Nebraska, University of  73
Norris, F. W.  170
Nottingham Farm Institute  111

Oestrogens see hormones

Penicillin  47, 153, 155
Peppercorn, A. J.  58
Perrin, Michel  168
Phelps, Anthony  72
pigs  1, 2, 6, 19, 21, 36, 93, 110–113,
          118, 152–155, 157, 172,
          176–178, 180, 190–193, 195
          in America  112
          costing  192
          Danish piggeries  191
          'sweat box'  112, 191
          tail biting  113, 177
Pirie, Dr N. W.  200
Plant Committee Report on Fowl
          Pest  63
poultry  6, 18, 41, 42, 45–50,
          52–54, 56–68, 70–75,
          77–84, 108, 110, 134, 146,
          152–154, 156, 159–161, 163,
          164, 166, 168, 182, 184–186,
          194–199
          chronic respiratory
                    diseases  49, 152
          de-beaking  47, 71, 78, 120,
                    183, 196
          dubbing  82
          featherpecking and
                    cannibalism  48, 49,
                    67, 71, 120, 177
          fowl pest  50, 52, 53
          inspection of carcases  64
          manure as fertiliser  157, 198
          pecking order  48, 67, 174
          position in U.S.A.,
                    Netherlands,
                    Canada  63
          poultry keeping in
                    America  78
          specs  82, 120
          in U.S.A.  63, 78
          battery hens  36, 106, 109, 171,
                    176, 177
          in America  79, 163, 164
          Cage Layer Fatigue  74, 183
          costing  72
          in Denmark  105

Poultry (*continued*)
    broiler chickens  1, 6, 32, 41–55,
        58, 60, 106, 109, 118, 152,
        158, 164, 177
      in Georgia  41
      livers  77, 166
    poultry housing  48, 67, 72, 73, 194, 197
      deep litter  43, 67, 75
      slatted flooring  67
      'stimulighting' and
          'twilighting'  74, 81
      wire flooring  67
poultry packing stations  56–64, 184–186
Preston, Dr T. R.  35, 92, 108, 109, 192
Price, Dr Weston, 141
Protection of Animals Act 1911  5, 174,
    176, 178, 181, 188, 189, 191, 203
Protection of Birds Act 1954  6, 182


Quail  82, 106


Rabbits, broiler  109
rating of agricultural buildings  172
Reid, J. A.  66
Robinson, S. G.  45
Rowett Research Institute  7, 35, 92,
    107, 108, 192
R.S.P.C.A.  184, 186
Royal Veterinary Colleges  101
Roy, Dr J. H. B.  91, 92, 99, 100, 151, 176
rumination, prevention of *see* calves


Sainsbury, Dr D. W. B.  153, 192
Sanders, Dr  167
Scott, C. W.  50
Sellers, Dr K. C.  112
Shillam, Dr K. W. G.  102, 105
Sidley, Miss Dorothy *see* Humane
    Slaughter Association
Siller, Dr W. G.  74, 75
Sinclair, Dr Hugh  39, 171
Slats  113, 148–51
      for beef calves  89, 107, 177
      for veal calves  88–90, 93
slaughterhouses  63–64, 86

Smith, H. William  154
Smith, Mrs K. M.  49
Smith, William E.  162
Soil Association  146
stilbestrol *see* hormones
stress  15, 18, 23, 49, 76, 77, 111, 184
stunning, humane  59–61, 105,
    124, 185, 186
    Danish  61
subsidies  83, 150, 197, 198, 201


Taste in food  168
    synthetic  172
Thornber, Cyril  79
tranquillisers  38, 39, 108, 135, 155,
    182, 186
Tristan de Cunha  145
Trow, Edward  54
Trow-Smith, R.  88, 105, 193
turkeys  58, 106, 152, 156


Underdeveloped countries  199
U.S.A.  40, 41, 63, 68, 78
U.S. Department of Health,
    Education and Welfare  162
Unknown Vitamin  151


Veal *see* calves


Wales, University College of  179
Weeks, W. G. R.  41, 81
Wickenden, Leonard  39,
    159–163, 165
Williams, Dr Stephen  176, 189
Wiltshire Weights and Measures
    Department  157
Wilson J.  100, 101
Wisconsin, University of  68, 105
Wokes, Dr Frank  170
Wrench, Dr J. T.  140, 141
Wright, R. A.  59, 185, 186


Yudkin, Professor J.  169

# Publications Referred to and Quoted in the Text

*Adventures with a Naturalist* 151, 176, 183,
*Agriculture* 1, 2, 10, 13–16, 18–19, 24,
        32, 37, 38, 40, 44, 50–52, 55, 59,
        73, 76, 95–97, 102–105, 120, 145,
        147, 150, 157, 158, 163, 165, 167,
        176, 179, 181, 188, 192, 193, 197
*Animal Life* 190
*Bedfordshire Times and Standard* 43
*Bristol Evening World* 157
*The Broiler House* 44, 45, 53, 120
*Calf Rearing* 87, 88, 92, 94, 103, 188
*Chemicals in Food* 155, 156
*Conveyance of Live Poultry Order
        1919* 184
*The Countryman* 180
*Daily Express* 54, 108
*Daily Mail* 46, 107, 182
*Daily Telegraph* 50, 51
*East Anglian Daily Times* 183
*Essai de Caracterisation des Viands
        de Veau Insuffisantes*
*Evening Standard* 68, 152
*Experimental Husbandry No.5* 102
*Farmer and Stockbreeder* 35, 41, 45,
        58, 61–63, 66, 70, 71, 73–75,
        78, 81–83, 86–88, 92, 93, 95,
        103–105, 109, 111, 113, 119, 130,
        152–154, 156, 164, 165, 169,
        176–181, 189, 191–193, 195
*The Farmer's Weekly* 70, 138
*Farming Express* 42, 47, 48, 74, 75, 78,
        81, 83, 85, 107, 152, 164, 189, 192
*The Financial Times* 43, 107, 108, 167, 169
*The Haughley Experiment* 146, 147
*Hazards of Life* 169
*Hen Batteries* 77, 181, 184
*Humane Slaughter Association Annual
        Report* 59
*Husbandry and Diseases of Calves* 90, 188

*Incubation and Hatchery Practice* 76, 172,
*King Solomon's Ring* 48
*The Lancet* 171
*May and Baker Laboratory Bulletin* 182
*The Medical Testament* 146
*Mother Earth* 168
*News Chronicle* 45, 93
*Nutrition and Health* 140
*Nutrition and Physical
        Degeneration* 141–142
*The Observer* 23, 37, 86, 106, 124
*Our Daily Poison* 39, 65
*Our Synthetic Environment* 154, 155, 157
*Oxford Mail* 65, 66
*Plant Report* 46
*Poultry World* 46, 47, 49, 53, 54, 57, 61,
        66, 70–72, 75, 77–80, 82, 84, 110,
        152, 160, 164, 168, 185, 198, 199
*Protection of Animals Act* 1911 5, 174,
        176, 178, 181, 188, 189, 191, 203
*Protection of Birds Act* 6
*The Scientific Basis of Kindness to
        Animals* 174
*Scientific Principles of Feeding Farm
        Livestock* 91, 99
*Silent Spring* 3, 149, 150
*Second Conference on the Health of
        Executives* 39
*The Smallholder* 48, 70
*Sunday Times* 39, 200
*Third World Food Survey* 181
*The Times* 51, 96, 105, 163, 190
*U.F.A.W. Courier* 86, 92, 104
*Veterinary Dietetics* 100
*The Veterinary Record* 38, 62, 87, 89, 97,
        100, 110, 187
*The Wheel of Health* 140
*Wiltshire Gazette and Herald* 50
*Yorkshire Post* 186